D0324385

GRAPHING DATA

Applied Social Research Methods Series
Volume 36

APPLIED SOCIAL RESEARCH METHODS SERIES

1. **SURVEY RESEARCH METHODS**
   Second Edition
   by FLOYD J. FOWLER, Jr.
2. **INTEGRATING RESEARCH**
   Second Edition
   by HARRIS M. COOPER
3. **METHODS FOR POLICY RESEARCH**
   by ANN MAJCHRZAK
4. **SECONDARY RESEARCH**
   Second Edition
   by DAVID W. STEWART
   and MICHAEL A. KAMINS
5. **CASE STUDY RESEARCH**
   Second Edition
   by ROBERT K. YIN
6. **META-ANALYTIC PROCEDURES FOR SOCIAL RESEARCH**
   Revised Edition
   by ROBERT ROSENTHAL
7. **TELEPHONE SURVEY METHODS**
   Second Edition
   by PAUL J. LAVRAKAS
8. **DIAGNOSING ORGANIZATIONS**
   Second Edition
   by MICHAEL I. HARRISON
9. **GROUP TECHNIQUES FOR IDEA BUILDING**
   Second Edition
   by CARL M. MOORE
10. **NEED ANALYSIS**
    by JACK McKILLIP
11. **LINKING AUDITING AND METAEVALUATION**
    by THOMAS A. SCHWANDT
    and EDWARD S. HALPERN
12. **ETHICS AND VALUES IN APPLIED SOCIAL RESEARCH**
    by ALLAN J. KIMMEL
13. **ON TIME AND METHOD**
    by JANICE R. KELLY
    and JOSEPH E. McGRATH
14. **RESEARCH IN HEALTH CARE SETTINGS**
    by KATHLEEN E. GRADY
    and BARBARA STRUDLER WALLSTON
15. **PARTICIPANT OBSERVATION**
    by DANNY L. JORGENSEN
16. **INTERPRETIVE INTERACTIONISM**
    by NORMAN K. DENZIN
17. **ETHNOGRAPHY**
    by DAVID M. FETTERMAN
18. **STANDARDIZED SURVEY INTERVIEWING**
    by FLOYD J. FOWLER, Jr.
    and THOMAS W. MANGIONE
19. **PRODUCTIVITY MEASUREMENT**
    by ROBERT O. BRINKERHOFF
    and DENNIS E. DRESSLER
20. **FOCUS GROUPS**
    by DAVID W. STEWART
    and PREM N. SHAMDASANI
21. **PRACTICAL SAMPLING**
    by GARY T. HENRY
22. **DECISION RESEARCH**
    by JOHN S. CARROLL
    and ERIC J. JOHNSON
23. **RESEARCH WITH HISPANIC POPULATIONS**
    by GERARDO MARIN
    and BARBARA VANOSS MARIN
24. **INTERNAL EVALUATION**
    by ARNOLD J. LOVE
25. **COMPUTER SIMULATION APPLICATIONS**
    by MARCIA LYNN WHICKER
    and LEE SIGELMAN
26. **SCALE DEVELOPMENT**
    by ROBERT F. DeVELLIS
27. **STUDYING FAMILIES**
    by ANNE P. COPELAND
    and KATHLEEN M. WHITE
28. **EVENT HISTORY ANALYSIS**
    by KAZUO YAMAGUCHI
29. **RESEARCH IN EDUCATIONAL SETTINGS**
    by GEOFFREY MARUYAMA
    and STANLEY DENO
30. **RESEARCHING PERSONS WITH MENTAL ILLNESS**
    by ROSALIND J. DWORKIN
31. **PLANNING ETHICALLY RESPONSIBLE RESEARCH**
    by JOAN E. SIEBER
32. **APPLIED RESEARCH DESIGN**
    by TERRY E. HEDRICK,
    LEONARD BICKMAN, and
    DEBRA J. ROG
33. **DOING URBAN RESEARCH**
    by GREGORY D. ANDRANOVICH
    and GERRY RIPOSA
34. **APPLICATIONS OF CASE STUDY RESEARCH**
    by ROBERT K. YIN
35. **INTRODUCTION TO FACET THEORY**
    by SAMUEL SHYE
    and DOV ELIZUR
    with MICHAEL HOFFMAN
36. **GRAPHING DATA**
    by GARY T. HENRY
37. **RESEARCH METHODS IN SPECIAL EDUCATION**
    by DONNA M. MERTENS
    and JOHN A. McLAUGHLIN

GRAPHING DATA

Techniques for Display and Analysis

Gary T. Henry

Applied Social Research Methods Series
Volume 36

SAGE Publications
International Educational and Professional Publisher
Thousand Oaks London New Delhi

for my parents, Otis D. Henry and Marcella Tyler Henry

For information address:

SAGE Publications, Inc.
2455 Teller Road
Thousand Oaks, California 91320

SAGE Publications Ltd.
6 Bonhill Street
London EC2A 4PU
United Kingdom

SAGE Publications India Pvt. Ltd.
M-32 Market
Greater Kailash I
New Delhi 110 048 India

Printed in the United States of America

Library of Congress Cataloging-in-Publication Data

Henry, Gary T.
Graphing data: techniques for display and analysis / author, Gary T. Henry.
p. cm. — (Applied social research methods series; vol. 36)
Includes bibliographical references and index.
ISBN 0-8039-5674-6. — ISBN 0-8039-5675-4 (pbk.)
1. Statistics—Graphic methods. I. Title. II. Series: Applied social research methods series; v. 36.
QA276.3.H46 1995
001.4′226—dc20 94-26841

95 96 97 98 99 10 9 8 7 6 5 4 3 2 1

Sage Project Editor: Susan McElroy

Contents

Acknowledgments

Graphing Data and the ideas that are used in it have been influenced by many people who stimulated my thinking, critiqued my ideas, and labored with me in devising graphical solutions to data problems. While mounds of intellectual debt from other authors pile up in the references, more direct, personal influence can get short shrift. My original interest in using graphs to communicate data came from a seminar taught by Ed Tufte that I chanced to take in Michigan during the summer of 1979. Two graphic artists, David Porter and Valerie Murphy, worked with me in putting Tufte's ideas and some of our own to practical use in applied research reports. Mary Stutzman and Virginia Hettinger provided meticulous reviews that raised questions that were usually followed with suggestions that made the book more useful. Along with an anonymous reviewer, these two reviews greatly assisted me in maintaining the focus of the book. Debra Rog and Len Bickman helped me craft a book that presents theories and new ideas that relate to data graphics in a way that can be used by those who do applied research. The ideas would not have been tested so often or used so much without the spirited interplay that has taken place at the Center as we have worked on data analysis and reports. Leo Simonetta, Kevin Kitchens, and Chris Craver lent their special talents to pushing the limits of graphical software and devising ways to produce yet another variation of a new graphical display.

1

Applied Research and the Use of Graphical Displays

We find graphs in almost every nonfiction publication that we pick up these days. From newspapers and magazines to technical journals, authors and editors infuse their writing with graphs and charts. Graphs, such as trend lines and bar charts, can set the stage for a discussion, convey a message, or reinforce a central point. Data-based, analytical graphs are essential tools of the sciences, and for the technically trained, they often convey data in a concise and readily digestible format. In Figures 1.1a and 1.1b, two types of data displays are contrasted: the first is a simple pie chart reproduced from a newspaper article; the second displays the relationship between two variables from a study of students of traditional college age and nontraditional students. From the first graph, we observe that over half of the public believes Olympic athletes should not be paid. Figure 1.1b is more complex and interesting, showing that spouse's attitudes about attending college are more important to married students in the traditional age group (< 25) than to older nontraditional students. Each student is represented by a filled circle on the display and the mean for each group is represented by the line. The graph allows the viewer to see the distribution of responses as well as the mean in one quick scan.

The use of graphical data displays such as these has not always been so common. Only 10 years ago, a publication-quality graph required a graphic artist or draftsperson to transfer a data set onto the page of a report. The process consumed considerable time and expense. Now, most software programs that process quantitative data will produce graphs. Spreadsheet software, such as Lotus 1-2-3 and Excel; statistical software for mainframes and personal computers, such as SAS, SPSS, and SYSTAT; and drawing and design software, such as Designer, DRAW, and Corel*Draw*, have the capacity to produce high-quality data-based graphics. These software programs and many others allow researchers to generate and refine graphical displays on the screens of their desktop computers.

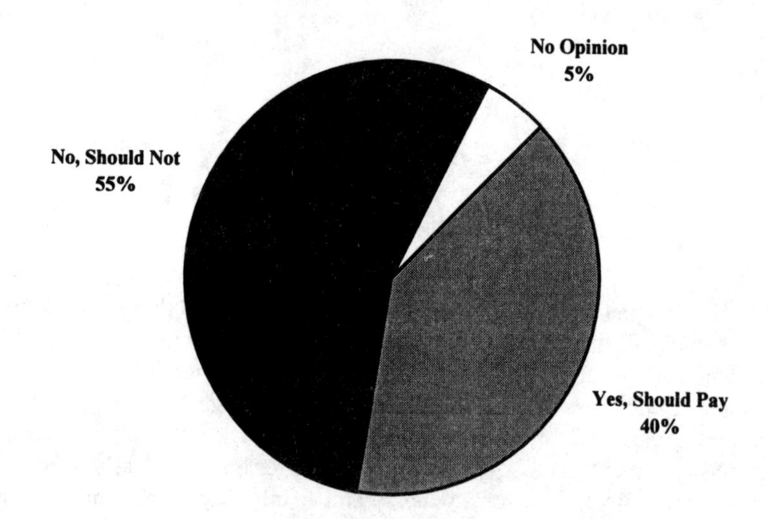

Figure 1.1a. Georgians' Opinion About Paying Olympic Athletes
SOURCE: Data from *Atlanta Journal-Constitution*/Georgia State University Poll (1993). Reprinted with permission from *The Atlanta Journal* and *The Atlanta Constitution*.

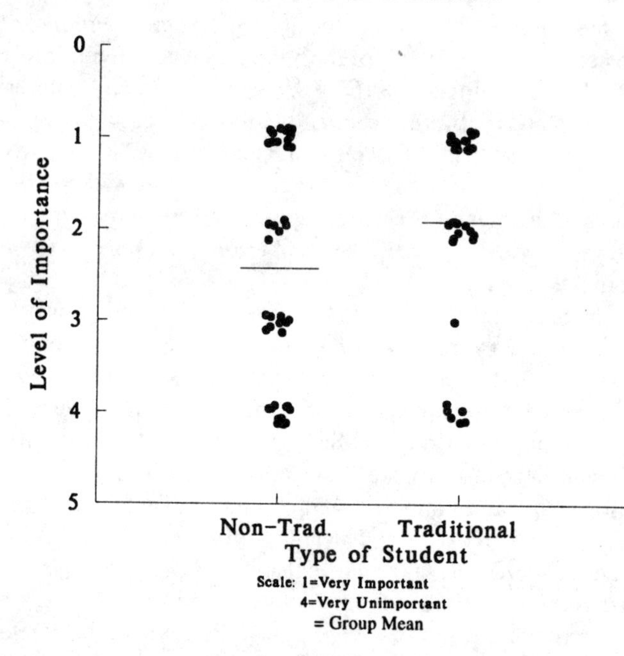

Figure 1.1b. Importance of Spouse's Attitude About Attending College to Students of Traditional and Nontraditional Age

It is fortunate that the means for producing graphical displays has come at a time when we are coming to appreciate their power and utility. Throughout this book I will be concerned with communicating data by producing graphical displays of quantitative data. The terms *graph, graphical display, chart,* and *graphic* will be used more or less interchangeably. All these terms refer to graphs that are based on quantitative data and have the purpose of visually displaying data. The power of graphs comes from their ability to convey data directly to the viewer. In this sense, graphs reveal data to viewers. Viewers use their spatial intelligence to retrieve the data from a graph, a different source of intelligence from the language-based intelligence of prose and verbal presentations. The audience sees the data. Data become more credible and more convincing when the audience has a direct interaction with them. The communication process becomes more direct and immediate through graphical displays.

The power that stems from using graphs to communicate data directly with an audience can be thwarted, however. Badly designed or poorly executed graphs can obscure the data and try the patience of even a motivated audience. Most people have the ability to retrieve data from a well-composed graph: One education theorist goes so far as to state that this ability is "hardwired" into the brain (Wainer, 1992). Unfortunately, producing a good graph is not as easy as retrieving information from it, and in many cases, relying on the default graphics in some of the software programs cited above will not produce readily usable graphs. This book can guide the production of graphics that fulfill their communication purpose, something that I refer to as achieving graphical competence. In Chapter 2, I explain the concepts underlying graphical competence more fully, but first I will discuss the uses to which graphics are put.

USING GRAPHS: FROM DESCRIPTION TO ANALYSIS

Graphical displays of data function as descriptive information sources as well as analytical tools. Presentation graphics are frequently used in briefings and formal policy discussions of boards and elected and appointed officials. Presentation graphs allow the audience to visually access descriptive data while the speaker communicates verbally. Usually presentation graphs contain a limited amount of data, which allows

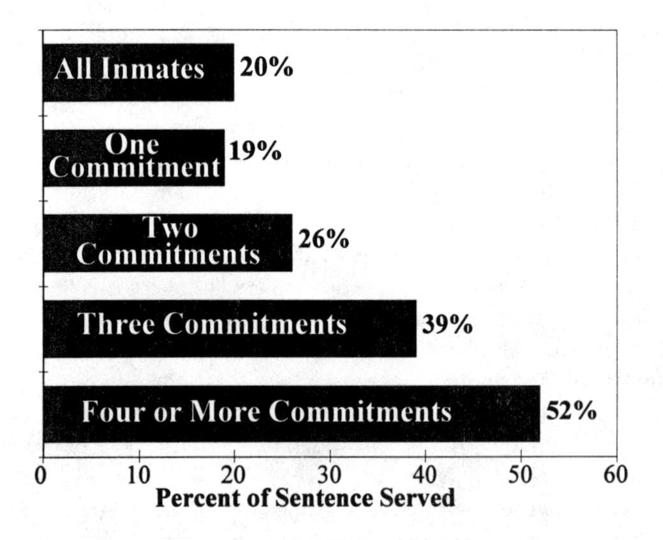

Figure 1.2. Sentences Served at Parole Eligibility
SOURCE: Recreated from graphic found in Joint Legislative Audit and Review Commission (1991, p. 3).

the audience to retrieve the message in a quick scan. Presentation graphs often introduce material or make a specific point. Figure 1.2 shows a good example of a simple, descriptive display.

Figure 1.2 shows a graphic the staff of a legislative oversight commission used to efficiently convey data on parole eligibility to legislators and the public: Inmates are serving a small fraction of their actual sentences, on average 20%, before being considered *eligible* for parole. For first-time offenders, this amounts to 19% of their sentence; those who have been to prison four or more times become eligible for parole after serving 52% of their sentence. A quick scan, independent of the text, communicates this basic information. The graph allows those in the audience to form their own conclusions about parole policies that lead to these results, and just as important, it invites other questions, such as: What is the percentage of the *sentence* actually served? Is the number of commitments the only cause of the variation in the percentage of sentence served before the inmate is eligible for parole? Further analysis of the graph leads to the conclusion that most of the inmates are serving their first commitment, as the average for all inmates is only slightly higher than for those serving their first commitment. This

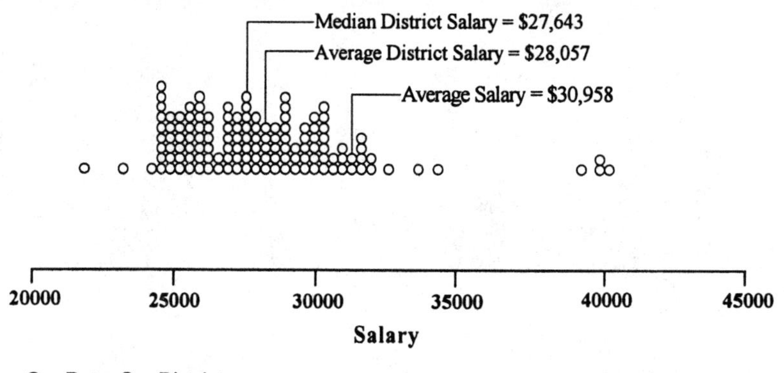

Figure 1.3. Average Teacher's Salary in Virginia
SOURCE: Data from Virginia Department of Education (1988).

observation requires substantially more analysis of the graph and an understanding of the way in which the overall mean is computed.

Data graphics should communicate information by facilitating the transfer of information from the researcher to the audience. And in turn, graphs should stimulate interest and enhance the understanding of the audience. Graphs, such as Figure 1.2, present a limited amount of data using a common type of graphical display, the bar chart. They can draw attention to a significant point, relying on spatial, rather than linguistic, intelligence to communicate data. But as we see in Figure 1.3, more can be done with graphics: They can be used for more detailed description as well as analytical purposes.

Figure 1.3 invites the audience to consider teachers' salaries in Virginia. The graphic depicts the average salary in each of the state's 133 school districts. Each dot represents the salary in one school district. The dots are stacked vertically to resemble bars in graphs of frequency distributions, sometimes referred to as histograms. The reader can quickly grasp the level and the variation of average salaries among the districts.

Several norms or averages are shown: the median of the 133 district averages, the average salary of the district averages, and the average salary in the state. The form of the graphic and the amount of data presented allows the audience to go beyond the information available from a scan and to describe the salary situation further. The audience can examine the relationship between each of the norms and between

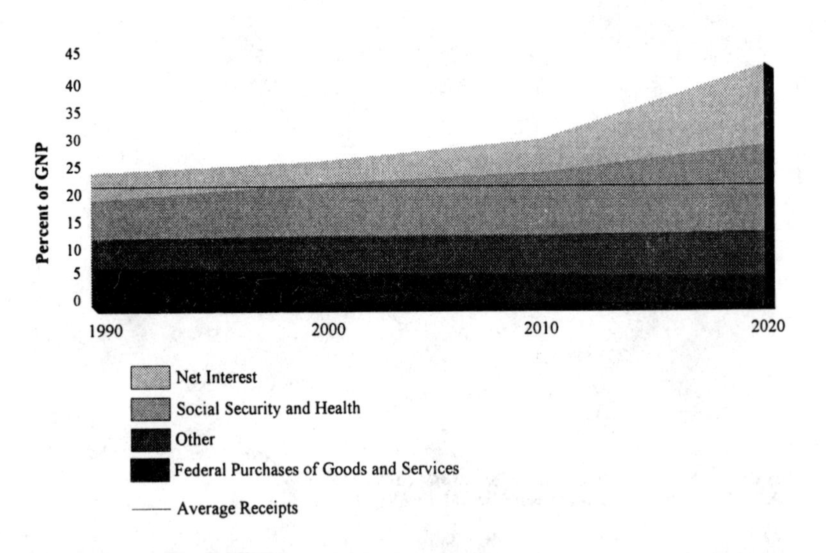

Figure 1.4. Federal Expenditures in the No Action Scenario
SOURCE: Reproduced from U.S. GAO (1992, p. 60).

the norm and the extremes. Only 13 districts pay more than the average salary. The average district salary is significantly higher because of these districts, indicating that they must be very large and employ many teachers. Twelve districts pay less than $25,000 on average. After a close examination of the graph, some may ponder why teachers' salaries vary so much across the state. In addition, viewers inside and outside the legislature may begin to ask what the state should use as the base for funding teachers' salaries. Should funding be based on the salaries actually paid to teachers and perhaps increase the base in districts that are wealthier and pay higher salaries? Or should the state use a norm, such as the median or the mean, for a base? Which norm would be better?

Good graphs encourage questions. However, some graphs hide more than they show. Sometimes the ability to communicate graphically is inhibited by trying to do too much in a single graph. In other instances, communication fails because the graphical designer is unsure of the purpose of the graph. One principal reason for the failure of a graph to communicate is improper execution of the basics of graphical design. Figure 1.4 is a reproduction of a graph that obscures the data and inhibits more detailed probing.

The purpose of Figure 1.4 was to raise concerns about the rapidly increasing federal deficit and motivate Congress to take action to reduce

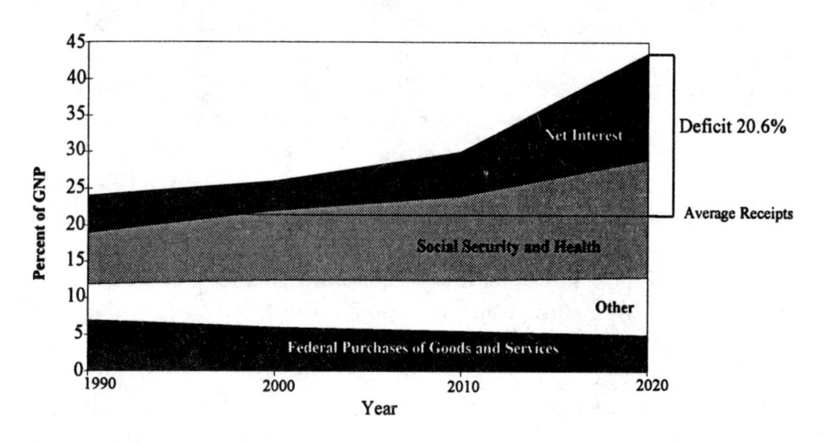

Figure 1.5. Federal Expenditures in the No Action Scenario
SOURCE: Recreated from U.S. GAO (1992. p. 60).

it. The text referring to the graphic states, "The deficit explodes to 20.6 percent of GNP by 2020, due in part to the projected dramatic rise in spending" (U.S. General Accounting Office [GAO], 1992, p. 60). Yet the graph appears anything but explosive and dramatic. Subtle differences in shading make the four trend lines, arranged in a cumulative fashion, difficult to distinguish. The scale of the graph—too wide for its height—stretches the display across the page and reduces the visual impact of the increase, thus contradicting the written message. The combined impact of the written and graphic communication is muddied.

The effect is reduced further in the original by the fact that the 20.6% figure does not directly appear on the graph, although with some effort, it can be obtained from the graph by subtracting the line that represents average receipts from the top of the cumulative line for the year 2020. The authors missed the potential to tie the graph to the text. A further problem arises from a failure to consider how the graph was to be incorporated into the report. The report, published in a bound, 8½-x-11-in. format, obscured the right side of the graph (the explosion), because that portion of the graph fell behind the curve of the created by the report binding. Figure 1.5 attempts to marry the graph with the text. The viewer now gets the impression that the growth in net interest payments produces most of the increase, which now appears more dramatic.

The three batches of data that have been presented graphically in Figures 1.2, 1.3, and 1.4 are principally useful for descriptive purposes. They show a summary of the data. They may be used to set the stage or to raise questions, but they seldom contain the means to answer questions. These graphs are located at one end of what Fienburg (1979) describes as a continuum of purpose for graphics. The continuum is anchored at one end by descriptive display of data and at the other end by analysis of data. Tukey (1988) points to two types of graphs with descriptive purposes: graphs that substitute for tables and graphs that show the result of some other technique. The trend line, which was found by Tufte (1983) to be the most popular graph, and the bar chart are two examples of graphs generally used for descriptive purposes.

The multivariate trends lines presented in Figure 1.5 assist in breaking the cumulative deficit increase into its component parts, which allows the viewer to begin to analyze the sources for the increasing deficit. This graph goes beyond description to encourage the reader to analyze the data contained in it. Tukey (1988) describes the graphs further toward the end of this continuum as "graphs to let us see what may be happening over and above what has already been described (analytical graphs)" (p. 38). The two-variable plot in Figure 1.6 moves further into the analytical realm.

Plotted in Panel A of Figure 1.6 are the percentages of students eligible for free or reduced-price lunch and students failing first-grade readiness tests by school districts. The panel shows the relationship between free lunch eligibility, which is a measure of poverty, and first-grade readiness at the school district level. Clearly there is a relationship between the percentage of children who are living in poverty in a school district and the percentage that fail at least one school readiness exam. The relationship appears to be strong, positive, and linear. Although we can quibble about the measures, it seems clear that school districts with high proportions of disadvantaged students have ground to make up compared to other districts. One could reasonably conclude that the difficulty of educating students increases in proportion to the amount of poverty in a school district.

In Panel B, a line drawn using standard linear (least squares) regression and summarizing the relationship between the level of poverty and the percentage of entry test failures has been added. The line adds the "norm," or the predicted percentage of entry test failures for each level of students living in poverty. Information about the line could be communicated by listing the coefficient estimates from the regression

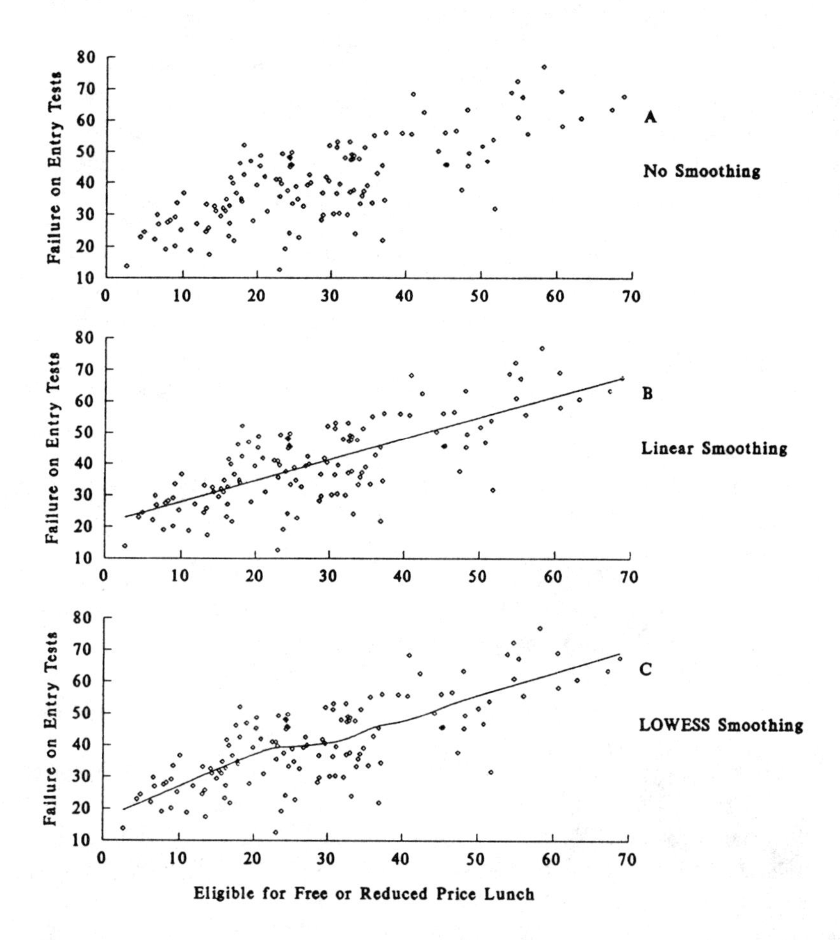

Figure 1.6. School District Percentage of Students Eligible for Free Lunch and Entry Test Failures

output. But the graph with the line is much more compelling and quickly assimilated, especially by a lay audience. The graph also retains the visual sense of variation in the level of poverty and failure of test scores in the districts and variation from the "norm." In addition, the graph allows one's attention to be drawn to the three cases that appear to have very low entry test failures for the level of poverty among students' families. These three districts, with levels of poverty at about 23%,

38%, and 51% and percentages of failures well below the line, appear to be unusual, or outliers. Outlier analysis, or the identification of unusual or anomalous cases, is something that is readily done with graphs such as these.

Panel C introduces an analytical procedure that is uniquely graphical. The line drawn through this graph is computed by a smoothing technique called LOWESS smoothing, developed by Cleveland (1979) and described in Chapter 6 of this book. The LOWESS line is straight when the relationship is linear, and thus allows the researcher to assess whether the relationship between two variables is linear. In this case, the assumption of linearity required for ordinary least squares regression is supported. This type of diagnostic analysis is most readily performed by graphical means.

These three panels illustrate graphs located more toward the analytical end of the continuum. However, it is not only the form of the graph that determines its place on the continuum. The choice of a graphical form can limit the sophistication of the graphical information displayed: For example, pie charts are only capable of breaking a whole into its component parts, whereas bivariate plots, such as those in Figure 1.6, can show relationships between two variables, the "norm" for the relationship, and outliers. Together with the graphical form, the amount and type of information displayed affects the readers' ability to penetrate the data further.

Most graphs in popular media present limited amounts of information. The bar chart and the trend graph are examples of graphical forms that usually present limited amounts of data. The bar chart displaying percentage of sentences served at the time of parole eligibility in Figure 1.2 falls into that category. The data displayed convey a quick summary of the situation, in this case the amount of time served before parole eligibility, but they also encourage a variety of questions that can provide a point of departure to lead the reader to other important data. Graphs such as those in Figures 1.2 and 1.3 are important for laying the descriptive groundwork in a research presentation, report, or article. They will be discussed in Chapters 3 and 5, respectively.

Analytical graphs combine the attributes of presenting more data and using a form that reveals the data to the reader. Analytical displays "encourage the eye to compare different pieces of data" and "reveal the data at several levels of detail, from a broad overview to the fine structure" (Tufte, 1983, p. 13). Time series graphs such as Figure 1.5 have some analytic capacity, especially when they include markers for significant events or multivariate trends. But the greatest analytic potential is available with graphical formats that present data in more abstract forms, such as the bivariate plot in Figure 1.6.

TAILORING A GRAPH TO ITS PURPOSE

The most basic dimension concerning the purpose of a graph is the analytic-to-descriptive use of the display. As the previous section has illustrated, graphs can be arrayed along a continuum from descriptive to analytical. Other purposes for graphical displays raise important considerations for designing and executing the displays. Motivating the audience to access the data is another important purpose for graphs. A graph can break up extensive narrative and encourage the reader to continue to engage the material. Sources of information that rely on sales and circulation to stay in business, such as newspapers and magazines, use graphical displays extensively. Their publishers make substantial investments to harness the power of graphics.

Tufte (1983) cites another purpose for graphical displays that bears on motivation of the audience: to decorate the text. It is obvious that popular publications as diverse as *USA Today* and *The Economist* have not overlooked the decorative and entertainment value of graphics. They include graphs in every issue. Although decoration may seem an anathema to some researchers, if it motivates the audience to examine important data, it is worthwhile.

The motivational and decorative aspects of graphical displays have a special importance for applied researchers. Hedrick, Bickman, and Rog (1993) provide an apt description of applied research as using "scientific methodology to develop information aimed at clarifying or confronting an immediate social problem. Its environment is often a messy one, with pressures for quick and conclusive answers, sometimes in political contexts" (p. 2). Applied researchers supply information about important social problems to a variety of audiences. The label *applied researcher* covers numerous types of professionals, including evaluators, educational researchers, policy analysts, legislative oversight committee staff, and medical researchers concerned with patient compliance.

The information provided by applied researchers is important for the professionals who deal with problems. In many cases the public also must comprehend and act upon such information. Applied research must reach audiences often busy with workaday issues and untrained in the use of data. The information must be presented in a form that motivates the audience to attend to it; findings rather than methods are their first priority. Applied researchers must compete for audience time with information generated by the media and therefore must adopt broader entertainment values, but without sacrificing credibility or giving way to bias.

Graphics are central in enabling applied researchers to communicate with audiences. Graphs allow applied researchers to directly convey their findings and the meaning of the findings. These tasks are descriptive, as in the first two graphical examples in this chapter. Researchers must describe the problem, provide context, and convey the importance of the aspect of the problem that they will address. They must persuade the audience of the importance of their topic and approach. Figure 1.2 displays a finding that establishes the context for the rest of a study of the parole system. It alerts legislators and the public that sentences may not mean what the audience expected: the graph attempts to dispel the myth that sentence length determines parole eligibility.

Researchers must both convince audiences of their analysis and allow audiences to analyze the data and determine the meaning for themselves. By showing the teachers' average salary for each school district in the state, the question of choosing a base for funding is clarified. The problem is made credible and real because of the data presentation. All school districts, including those in each legislator's district, are displayed. The use of one symbol for each district removes the abstract nature of the histogram and grounds the graph. Each district can be found on the graph. Some may receive more from the state than they are paying for teachers' salaries, others less. The problem is set by the visual analysis of the data: the question, though technical, is made plain and awaits a rationale and a solution.

These tasks, describing and analyzing data, are more difficult because the main audience for applied research is usually not other researchers. Basic research finds its main audience in the same research profession and specialty as the author. Writing for basic research is therefore more terse and technical. More can be assumed by the author in terms of audience knowledge of the subject, prior training, and familiarity with the way data are presented. For applied researchers the problem is completely different. For example, applied researchers in the field of education may have teachers, administrators, lay school board members, parents, and taxpayers as their main audience. Terse and technical communication can be seen as purposely obscuring "important information." It can cause the researcher to lose an audience and therefore lose the potential for using information to resolve a pressing problem.

The tasks of (a) describing the problem at hand and the point of departure for the study, then (b) displaying the findings are two significant parts of reporting applied research. Experienced applied researchers will immediately see the structure of their reports and papers in these two tasks (they may also notice the absence of another ubiquitous—and often tedious—feature of applied research reports, the description of the method). The first two tasks are well suited for graphical display. Many applied researchers use graphics in their reports, such as the examples shown in this chapter. But overall, despite their utility in communicating with nontechnical audiences, graphical displays of data may be underutilized.

In this book, I attempt to aid the applied researcher who wishes to mobilize the graphical displays as tactical supports in the communication of research data. Why applied researchers? For one thing, their work has direct relevance for large audiences. And for another, their work could better illuminate public policy discussions if it were more widely communicated and more broadly understood.

THE PLAN FOR THE BOOK

The purposes of this book are to encourage greater use of graphical displays in applied research and evaluation, and to provide applied researchers with guidance for graphing data. For these purposes, I believe that this book is both urgent and timely. Urgency arises from applied researchers having important information to add to the pressing issues of the day, information that is not always understood or accepted by the public. An urban scholar convinced of the need for social scientists to play a role in addressing the current urban crisis and of our inability to convey findings to the public has called for a new type of journalist to translate our work (Marris, 1992, p. A40). I am convinced that we who are social scientists must take the presentation of data more seriously ourselves and not depend on outsiders to do this for us. With competent graphical displays, we can do much of the work.

The book is also timely because we now have both data and the software necessary for graphing that data and incorporating the graphs into text. For example, every three months the Survey Research Lab at Georgia State University conducts a statewide poll. Within 48 hours of completing the telephone interviews, the staff produce a topical report on some of the major findings, which is made available to the press. The report includes statistical analysis and comments from academic researchers and a variety of graphical displays that are picked up and used in the media. A recent report featured a graph displaying narrow lines to show the confidence limits for the estimates depicted on the ends of bars (Cleveland, 1984; Wilkinson, 1990) to encourage the public and journalists to understand that the estimates, represented by the endpoints of the bars, have a range around them which 19 times out of 20 includes the true value. (See Chapter 2, Figure 2.5, for an example.) High-quality graphics displaying highly salient data can be produced quickly when interest, time, and data are combined with the appropriate software. Some software programs—in the example above the software used is SYSTAT—allow researchers to go beyond conventional graphical formats and stretch their audiences' knowledge.

As I have already pointed out, many statistical and spreadsheet software programs will turn out report-quality graphics. Most word-processing software will incorporate graphics into the text and some allows the analyst to enhance the presentation quality of the graphs. Certainly, this capacity exists with all popular desktop publishing software. If the data we collect are important enough for us to labor over the sampling, the data collection, and the data analysis, doesn't it justify using available tools to inform the public? If you agree, then the remainder of the book will give you insight into some good practices to be followed.

In Chapter 2, I will lay out a framework for what we know about perception that relates to graphing data and use of graphical information. In this chapter, graphical competence will be described. The term *graphical competence* is used to connote the complex production-consumption relationship between the maker of the graph and the one who views it. Elements that have been pieced together from research on visual perception and *graphicacy,* that is, the ability to retrieve information from graphics, will be discussed. Their practical implications for producing graphs will be presented in detail.

Chapters 3 through 7 are organized around data summarization, analysis, and presentation issues that we find in applied research. We will begin with summarizing data in Chapter 3. Summarizing recalls the descriptive tasks that I discussed above: describing the issue or problem, providing context, and establishing a point of departure for the study. This chapter deals with common graphical formats, such as bar charts and pie charts. In Chapter 4, a situation common in applied research and evaluation is engaged: displays of multiple units. All too often, the only recourse for providing data on multiple units, be they states, schools, or program sites, is a table. Graphical alternatives to tables are offered in this chapter, as well as some means for improving tabular displays.

Both Chapters 5 and 6 concern bivariate plots. Chapter 5 addresses trend data, where one variable is always some measure of time. Because of the predominance of trend graphics in popular as well as scientific publications, trends require a chapter in their own right. In Chapter 6, I discuss two-variable scatterplots as graphical support for correlation and regression. This chapter will stress designing and using graphical supplements for these common analytical methods. Chapter 7 illustrates graphical methods for displaying data from *t*-tests and analysis of variance. These methods emphasize the variability of data and are used to emphasize the need to be aware of the magnitude of differences and the statistical significance of the differences by showing variation and the summary statistics together.

I also include some principles of competent graphical design applying to graphical types not included in this book. These principles are sprinkled throughout Chapters 3 to 7. A summary of the guiding principles as well as some general comments on the finish elements of displays, such as titles and legends, is included in Chapter 8. Principles such as these are best learned in the context of a specific example. Therefore, the principle involved with the graphical example will be pointed out when the graph is discussed. These are principles that, by and large, I have found in the work of others, such as Tufte, Cleveland, Tukey, and Wainer, and use in my own work. Chapter 8 will close the book with some ideas concerning aesthetics and communicating information in a democracy in relation to graphical displays.

As you can see, the book is organized around the needs of applied researchers. It gives data primacy. The data are the starting point for applied researchers. Purpose and then graphical type are placed into the organization. Finally, the principles and options that apply to specific design choices are provided. Chart 1.1 provides a guide for the researcher who approaches the book data in hand, ready to graph.

Not all types of graphical displays are featured in this book. I have concentrated on the types that in my experience as a dyed-in-the-wool user of data in policy, administrative, and academic positions, I have found most useful. Many graphical displays such as the Gini curve or the "isthmus of acceptance" for analyzing test items (Wainer, 1991) are

Chart 1.1
Quick Reference for Using *Graphing Data*

Type of Data	*Purpose*	*Graph Type*	*Chapter*
Parts of whole (% & proportions)	Descriptive	Bar chart	3
		Pie chart	3
		Divided bar chart	3
Univariate distributions	Descriptive/ Analytical	Frequency bar	3
Trends			
1 variable, many periods	Descriptive	Time series plot	5
multivariate, 2 periods	Descriptive	Time series plot	5
multivariate, many periods	Descriptive	Time series plot	5
Individual Cases			
univariate	Descriptive	Bar chart	3/4
multivariate	Analytical	STAR, profiles	4
single case analysis	Analytical	Box plot, stem & leaf, dot charts	4
Two variable relationships			
symmetric/correlation	Analytical/ Descriptive	Scatterplot	6
asymmetric/regression	Analytical/ Descriptive	Scatterplot/slope	6
asymmetric/regression	Analytical/ Descriptive	Residual plot	6
asymmetric/ANOVA	Analytical	Categorical plot	7
Multivariate relationships			
symmetric	Analytical	Scatterplot matrix	6
symmetric	Analytical	Cluster scatterplot	6
asymmetric	Analytical	3-variable plot	6

simply too specialized for this type of book. In addition, any methods that require multiple colors or contours have not been included. I assume that you will be as bound to the plain paper copier and transparency projector as I am. I hope that we can be sufficiently successful within these limitations that we can raise our expectations, in ways that I mention in the last chapter, in the near future. Finally, I have avoided a few graphics that I consider interesting, but without practical application for applied research. For example, I find Chernoff faces (1973) an interesting way to present multivariate, multi-unit data, but I doubt that many legislators, administrators, or members of the public would spend much time picking out differences between two cartoonlike faces when trying to understand differences in two program units.

EXERCISES

1. For one week, clip the articles from your local newspaper that have graphs that display data. Review the text that accompanies the graphs. Pick out the best example of a graph that illustrates the text. Analyze the graph. Could it stand alone and make the point without the text? Explain how the graph visually depicts the information.

2. From the same collection of graphs, pick out the most interesting graph. What brought the graph to your attention? Write down the information that the graph communicates to you. What questions do you want to ask after viewing the graph?

3. Pick up a national weekly news magazine and examine the articles that use graphs. How do these graphs compare to the newspaper graphs?

4. Review one issue of a journal in your field for the graphics that are used. How many graphs are used? Are they descriptive or analytical? Do the graphs appear to contain more or less information than the popular graphs? Describe the clarity of titles and labels, purpose of the graphs, and visual appeal.

5. Carefully examine Figures 1.4 and 1.5. What are the differences between the two?

2

Graphical Competence

From the start, graphing presents a difficulty. We must display a facet of reality, some bits of meaningful data, on a flat piece of paper, a transparency, or a computer monitor. The data are rich, yet our graphical resources are limited to a single plane and our creativity in using marks on the plane to transcend two dimensions. Tufte (1990) poses the problem most eloquently: "The world is complex, dynamic, multidimensional; the paper is static, flat. How are we to represent the rich visual world of experience and measurement on mere flatland?" (p. 9). The problem is clear. We are stuck with an *X-Y* plane and must use the plane and ink to reveal the data. Yet the root of the problem is also the source of its greatest potential:

> visual perception has at its disposal three sensory variables which do not involve time: the variation of marks and the two dimensions of the plane. The sign-systems for the eye are, above all, spacial and atemporal. Hence their essential property: . . . spacial systems, graphics among them, communicate in the same instant the relationships among three variables. (Bertin, 1967/1983, p. 3)

Through graphical displays, the capacity exists to transmit data at the pace and interest level of the viewer. No other method of communication places such control in the hands of the receiver of information. Auditory systems, including the audiovisual combination of television, are temporally bound. Even the videodisc requires time to retrieve information and display it.

To create a full understanding of how to realize the potential of graphical displays, it is necessary to grasp the framework for "graphical competence." This framework ties together the various elements necessary to display and retrieve data accurately and with minimal effort and time. When this is done well, we can say that graphical competence has been achieved. Graphical competence results in the transmission of information from the researcher to the audience that views the graph. As shown by the framework of graphical competence, the transmission of information is analogous to a market transaction.

Chart 2.1
Graphical Competence: Achieving Graphical Competence

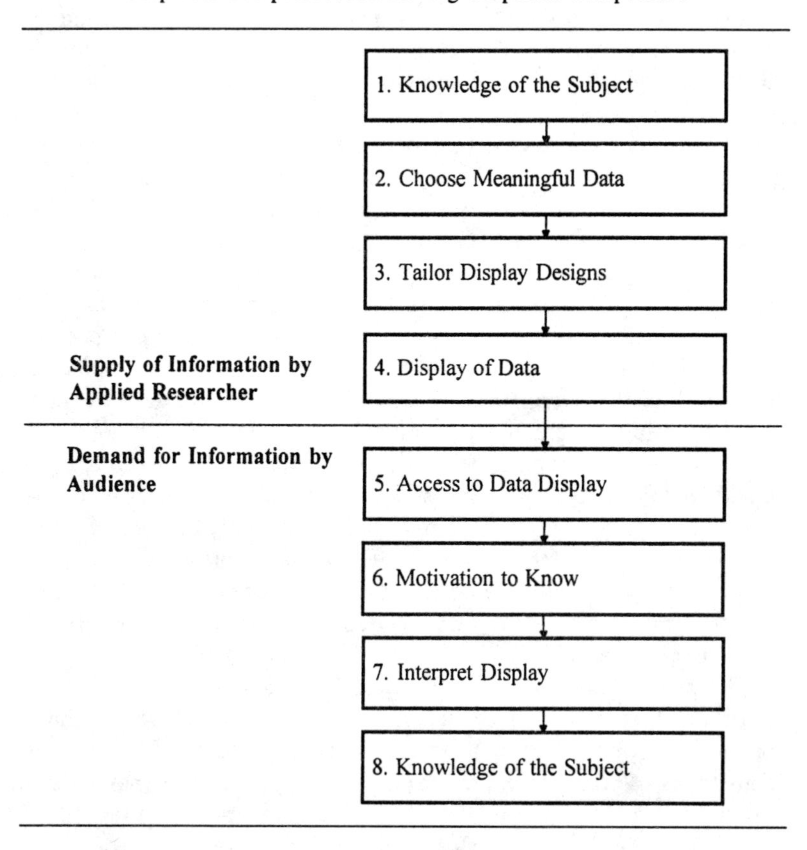

The researcher supplies graphical information and the audience consumes information.

Graphical competence, the process depicted in Chart 2.1, involves two distinct but connected processes—the production and consumption of graphical information. I will make extended use of the market analogy to better explain the roles of producer and consumer as well as the elements required by both to achieve a successful exchange of information. The market analogy is true to the concept that the process of transmitting information has two distinct and important roles—supplier (applied researcher) and consumer (audience for information). Both must be willing to undertake their role. This is as true for markets as it is for the transmission of data-based information. Too frequently, we focus on the technical aspects of graphic design or on the perception of displays without the broader framework that includes both the production and consumption of data displays. A very well designed and well drawn graph will not communicate the information it holds if no one is willing to take the time to view it. To achieve competence, a graph must not only be accurate, it must be examined.

The producer of the graph and the consumer each bring something to the table. The elements required of each are shown in Chart 2.1, where a series of linearly arranged milestones is shown by the arrows. To begin the process, graph makers must use their knowledge of the subject to select meaningful data, then design a display that reveals the data, and finally, execute the design accurately. Then the consumer becomes involved in the process. The first milestone for the audience is access to the display. The audience must be motivated to learn more about the topic by using the graphical data display; to continue with the market analogy, consumers must be sufficiently motivated to exhibit a "demand" for the information. As with the demand for other products, the demand for graphical information can be extant or it can be stimulated. The viewer must interpret the display, which involves tasks of visual perception and graphicacy. When graphical competency is achieved, researchers have used their knowledge of the subject to increase the knowledge of the consumer.

The process is depicted as linear in Chart 2.1, but this is a simple representation to emphasize the milestones involved in both roles in the process of communicating by means of graphical displays. In the end, consumers of graphical information have increased their knowledge of the subject of the graph and in the process often develop further questions. It is to be hoped that these questions motivate them to learn more and provide the researcher with the opportunity to present more information graphically. Thus, the process can become cyclical, with the researcher again assembling data and designing displays to present to the consumers. With this overview of the process, it is important to consider further the production and the consumption of graphical displays and the milestones that each must achieve. Achieving each of these milestones may require several iterations before it is complete. One of the goals of graphical competence—and this book—is to increase the base of knowledge about graphics and decrease both the number of iterations and the time required to communicate graphically.

The Audience for Applied Research Data: Consumers

The audience for applied research data should be broadly construed to allow for the widest scope of dissemination for the information in the displays. Following years of frustration with the public's and policy-makers' use of applied social research, researchers have spent much of their energy talking to one another and sporadically to the sponsors of their research and evaluations. The reluctance to expend the effort to go beyond the community of researchers and a part of the policy community should reconsidered. We should cast a larger net for the audience. Stakeholders in the research—program personnel, administrators, clients, advocacy groups—should be included in the audience (Rossi & Freeman, 1993, p. 423). But these groups represent only a beginning. The public, voters, or taxpayers should also be considered as members of the audience.

This conclusion follows from the value placed on an informed citizenry in the conceptualization of the democratic approach to evaluation devised by MacDonald (1976). Although MacDonald's (1976) work seems more oriented toward the community directly affected by an evaluation, the philosophy of democratic evaluation can be more inclusive, as the Jeffersonian notion of an informed citizenry would imply. Of course, this runs counter to the proprietary control over data that some research sponsors require. Yet it is not impossible or even improbable that we will find an audience interested in applied social research data, should we chose to cultivate it. Newspaper stands and the 24-hour broadcasts of CNN and C-SPAN belie our general sense of melancholy that people are not interested in information on social problems or public programs and policies. Most big-city dailies run social research data in graphical form in every section of the paper and, as Tufte (1983) found, the world press presents thousands of graphical displays over a 6-year period. What is lacking is "sophistication" in the choice of data presented and in the graphics used.

Public opinion polls are a form of applied social research that exemplifies the breadth of interest in data. Americans have literally become addicted to polls. Every day we hear about poll results from the media. One example of interesting poll data is presented in Figure 2.1, which shows Georgians' opinion of the quality of a range of public and private services (Poister & Henry, 1994). Contrary to the commonly held notion of the superiority of private services, the data show a wide range of

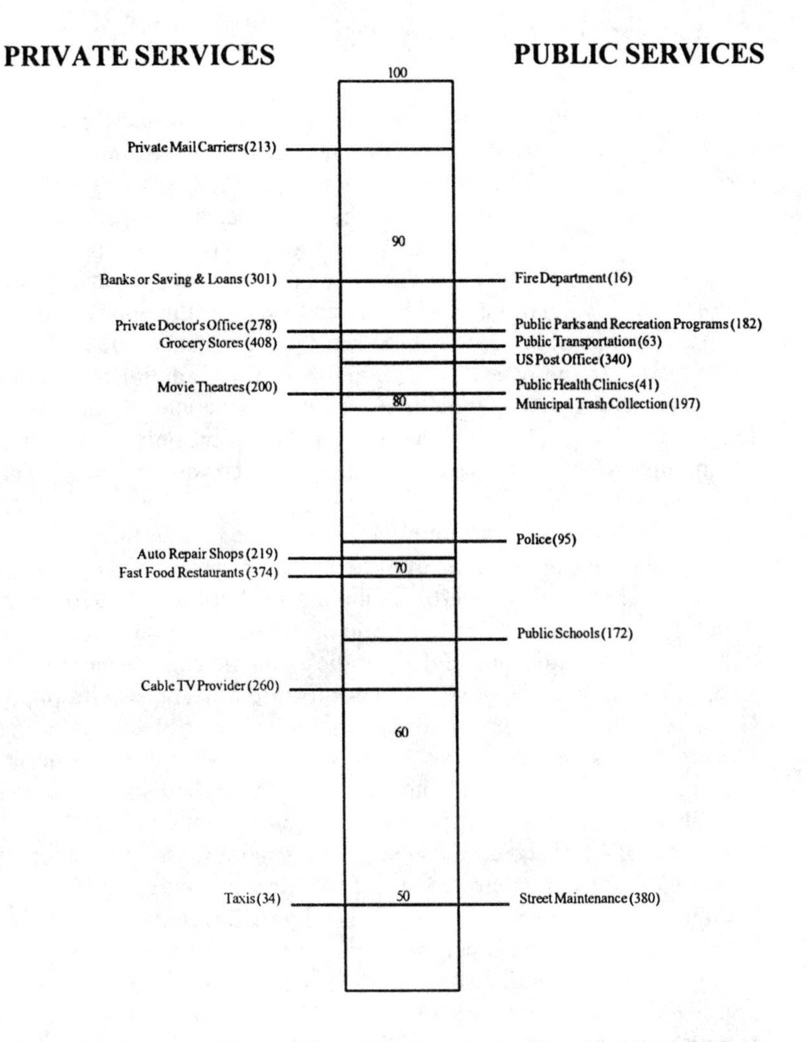

Figure 2.1. Percent of Recent Users Who Rate the Quality of Public and Private Services as "Good" or "Excellent" (number of recent users in parentheses)

variation in perceived quality of both public and private services with neither holding an overall edge. For example, while private mail carriers have the highest quality ratings by recent users, five private and six public services rate at the 80% good or excellent level. Such data expose myths and raise questions about the quality of the services themselves.

As further evidence of the public's interest in data we should consider the lengths to which the media go to obtain and report poll data. National and local media frequently sponsor polls and prominently publish their results. Why? Because poll data sells papers and circulation sells advertisements. Polls are so influential that we have a national debate and legislation concerning the reporting of exit poll data before the polls close. Public opinion data have achieved a place of prominence in the democratic processes in the United States, and in Great Britain are used to decide when to call parliamentary elections ("Britain this week," 1992)—and with the first breath of freedom in eastern Europe, we began to get reports from those countries' newly established polls.

Data presented in U.S. media are not limited to polls. Social research data quickly amassed in newsprint with the April 1992 riots following the verdict in the Rodney King beating case. Nearly every American saw the videotape made by a bystander of Los Angeles police force members beating King. When the police officers involved were acquitted, days of urban violence ensued. Figure 2.2 shows four of the eight graphical displays that were exhibited on the front page of the May 2, 1992, *Atlanta Journal-Constitution* following the disturbances in Atlanta. The media in this case were attempting to describe the extent to which blacks lag behind whites in economic well-being. They were proffering the argument that economic disparity might be the root cause of the spread of the violence. However, the two trends depicted in the graphs don't show the situation to have gotten worse recently, which does not support the overall argument of widening disparity as a cause of the riots. Although the choice of data may have been better informed by the research after the riots of the 1960s and early 1970s, the salient point for this discussion is that journalists looked to data for answers and felt the public was sufficiently interested in those data that newspapers should provide them.

Certainly, the data displayed in the media are not limited to social and public policy issues, but a great deal of systematically collected data are being presented. Also, data are being consumed by a large public audience, a public that pays for the information. Judging by the amount of data appearing in daily newspapers across the nation and on the 6 o'clock news, the public is deemed by profit-driven media executives to demand data. That data are in demand by the American public is not in question. Tufte (1983, 1990) has shown that the issues that deserve attention are which data should be presented and how they should be presented. These questions lead us to a consideration of the producers of information.

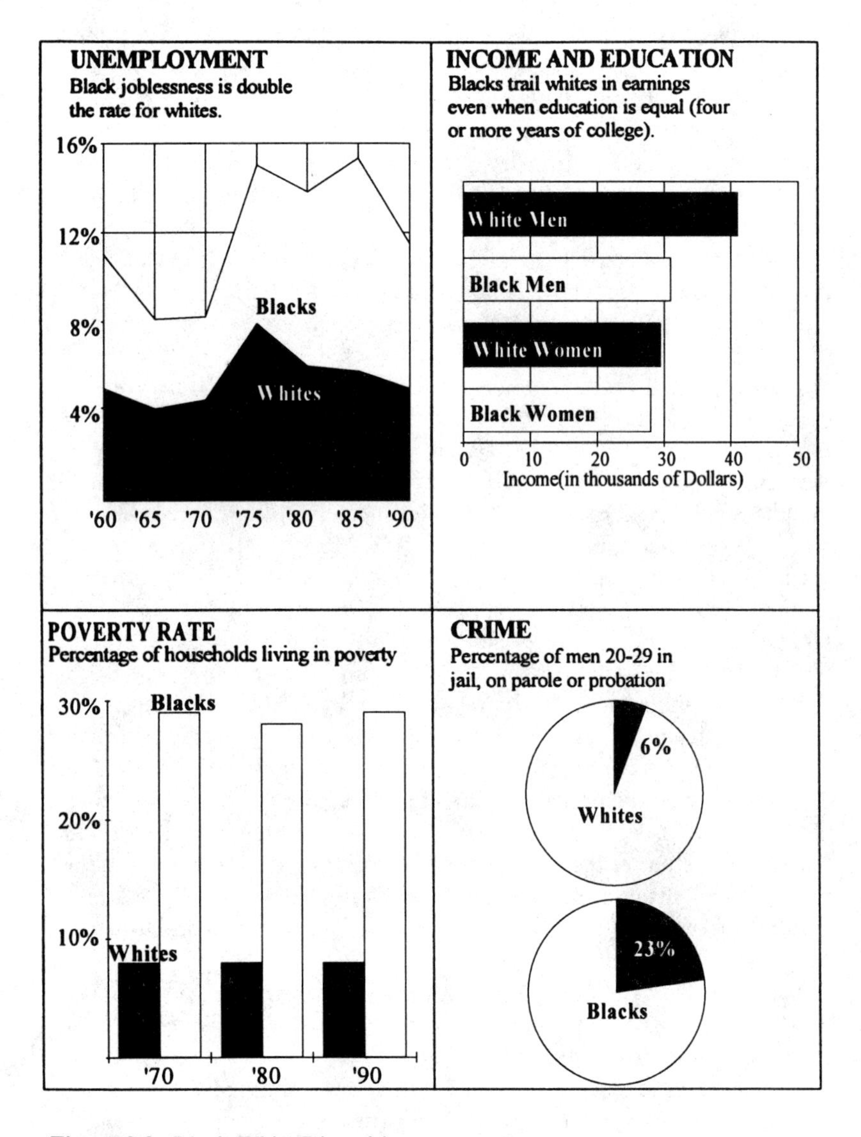

Figure 2.2. Black-White Disparities
SOURCE: "Black-White disparities," *Atlanta Journal-Constitution* (May 3, 1992, p. A1). Reprinted with permission from *The Atlanta Journal* and *The Atlanta Constitution*.

The Producers of Applied Research: Suppliers

From MacDonald's (1976) conceptualization of democratic evaluation, we can also begin to get a sense for who can supply the displays. MacDonald (1976) draws a line between research and evaluation, using the delineation that evaluators take on "real problems." This parallels the distinction between "messy" and complex social issues and the more narrowly scoped problems that arise in disciplinary investigations. The focus on "real problems" is useful, but using it to make a distinction between research and evaluation is not. Hedrick et al. (1993) use the same term to characterize the focus of applied research, which is more helpful for the purpose at hand. One finer distinction is needed, however, because basic research can also be directed at solving "real problems." For example, AIDS is a "real problem," yet much of the biological and biochemical research that is directed toward developing a cure or a preventative is not evaluation or applied research in the sense in which I use these terms here. It is discipline-oriented, empirical research that we typically label basic research. The *data* from research of this type are primarily to be communicated within professional and research communities to inform other research. Explanations, conclusions, and the day-to-day implications of the data concerning biological and chemical processes are often more broadly conveyed.

Applied social research, including evaluation, on the other hand, produce *data* of more direct relevance and use to audiences outside the professional and research communities. The concepts of social research have permeated society more thoroughly. Educational expenditures, test scores and dropout rates, crime statistics, inmate recidivism rates, product safety indices, and mass transit ridership rates are examples of data that contain important public information for citizens and taxpayers as well as those in the specialized professional and research communities. Rates of production of enzymes do not have the direct relevance to the public that these social data do, though they may be as important for the solution of a "real problem."

Both types of researchers may be dealing with "real problems." It is the distinction of directly transmitting *data* rather than conveying an *explanation of the data* that should be stressed here. It is the intent to convey data directly to audiences beyond specialized groups of professionals that will be used as the distinction for defining the suppliers of the type of data displays that I consider most directly in this book. I will refer to them as *applied social researchers.* The label is meant to refer to researchers in a wide variety of disciplines and whose subject matter can range from the efficiency of client intake procedures to consumer product evaluations to the change of political institutions to the incidence of AIDS.

THE SUPPLY OF GRAPHICAL DISPLAYS

In considering the supply of graphical displays, we must not lose sight of the fact that supply and demand are constantly providing feedback to one another. When the supply is too great or the quality is insufficient, the demand registered for the product may be low. Supply in excess of demand does not mean that the level of demand is fixed and a higher level cannot be stimulated; it means that the price is too high, the quality is too low, or the demand is not sufficiently elastic to accommodate the supply at the offered price. Price in this discussion is a function of time spent in retrieving data and other competing uses of time. How much time is the viewer willing to "spend"? The supplier must always consider whether or not the product can be offered at the price that the consumers will pay, whether sufficient demand can be stimulated, or whether the demand is cut short due to limitations on time.

The importance of time in processing data displays has long been examined in experiments on graphical forms. For example, since 1926 statisticians have been publishing papers that attempt to answer whether bar charts or pie charts are better icons for displaying percentages or proportions (Eells, 1926). Simkin and Hastie (1987) addressed the problem as one of both the accuracy with which the proportions were perceived and the time expended in making the judgments. They found that judgments of accuracy were no different for the simple bars and pie charts, but the pies took significantly more time to process. Thus, time is critical in establishing the preference for one type of graphical charts over another.

The main responsibilities of the applied social researcher in achieving graphical competence are collecting meaningful data, designing displays that reveal the data, and executing the design accurately. Achieving graphical competence places requirements on the researcher beyond the present standards of practice. Time, programming, and careful evaluation of graphs produced in the trial-and-error process of graphical display are required to achieve graphical competence. Also, producing competent graphs requires more than a little discipline from most researchers to focus attention on the data themselves rather on than the rigors of data collection or the artifacts of our analytical techniques. Tufte (1983) posits that excellent graphical displays should "induce the viewer to think about the substance rather than about methodology, graphic design, the technology of graphical production, or something else" (p. 91).

This requirement is not as simple as it may at first seem. A large portion of the credibility of our work resides in our choice of methods and the way in which they are executed. Information on methods often gets greater emphasis than substance in our papers and reports, in terms of volume and primacy of its presentation. For a wider dissemination of the data-based information, we must reverse the usual process. Data displays should represent the gateway to the research findings and only subsequently the methods. The displays should induce follow-up questions from the audience and review of the methodology to gain a fuller understanding of the phenomenon that is being reported on. The audience should be transformed from passive recipients of information to active learners. Viewers' interest shapes the manner of presentation and the ensuing discussion and stimulates greater investment in understanding the subject.

Meaningful Data

The determination of the meaningfulness of the data is a reasoned undertaking. The experience and expertise of the researcher with the topic guides the choices about which data are to be collected. Certainly, the researcher is principally responsible for making these decisions. Tradition, research sponsors, and other forces bear on the researcher's choices. Generally, a tradition of interest in a social measure offers a good guide for determining meaningfulness of data. Sometimes, however, tradition can inhibit a focus on the most meaningful data. For example, research on equity in education has traditionally focused on expenditures rather than outcomes of education. Most recognize that outcomes are more meaningful variables (Berne & Stiefel, 1984). The difficulties in collecting outcome data and in having an impact on outcomes are also well recognized.

Researchers' deliberations over what data are meaningful are certainly beyond the scope of this book, but one observation is critical. The data must be meaningful to both the researcher and the audience. "Meaningful" means interesting, important: "If the numbers are boring, then you've got the wrong numbers" (Tufte, 1990, p. 34). One goal of competent graphical displays is to get members of the audience to participate in the inquiry by viewing the data in the display and then asking their own questions, and when possible, viewing more data. The data must be sufficiently meaningful to the viewers that they are motivated to take an active role in the process. The exchange of meaningful data between researcher and audience is necessary to motivate the audience to look more deeply into the topic.

A significant factor in the determination of meaningful data is that the graph presents sufficient data to hold the interest of viewers and to draw them into the topic. Yet a graph that attempts too much, contains too much data, often ends up frustrating audience members. The amount of data that will work will be different for different viewers. The amount of data in a graphic can be calculated by counting the bits of information necessary to present the same information in a table (don't forget the labels). The number of data bits divided by the number of square inches in the graph is the data density or data per square inch (Tufte, 1983). An example of the calculation of data density can be done with Figure 2.3, which presents outcome measures for a school district and summary information for 14 school districts with similar characteristics. (See Henry, McTaggart, & McMillan, 1992, for details of the selection method.) The data density for this figure is 4.3 per square inch. The median data density in a variety of international popular and scientific publications is 8 per square inch (Tufte, 1983).

The data density in this example is reasonable for an audience of parents, teachers, and school board members. The data density is somewhat below the median for three reasons. First, the graphical format is unique to this data, and thus, the researcher can expect that the audience will need more time to process the information. Second, clear labels are important for each variable. In this case, each variable is an indicator of school district performance. Judgments and actions are being made based on this information—attendance and dropout rates indicate poor performance for the division for which these indicators are presented; test scores, for at least two tests, appear to be relatively high. Making sure labels are legible adds space to the graph. Third, the decision was made to summarize the data for the 14 benchmark school districts by presenting the median, 25th percentile, and 75th percentile. If the full data matrix from which these summary statistics were derived were plotted on the same graph, the resulting data density would be over 128 per square inch. The summaries seem more appropriate given the nature of the audience and the data. Support of the effectiveness of this graph was found in an experiment that determined that the intended users for

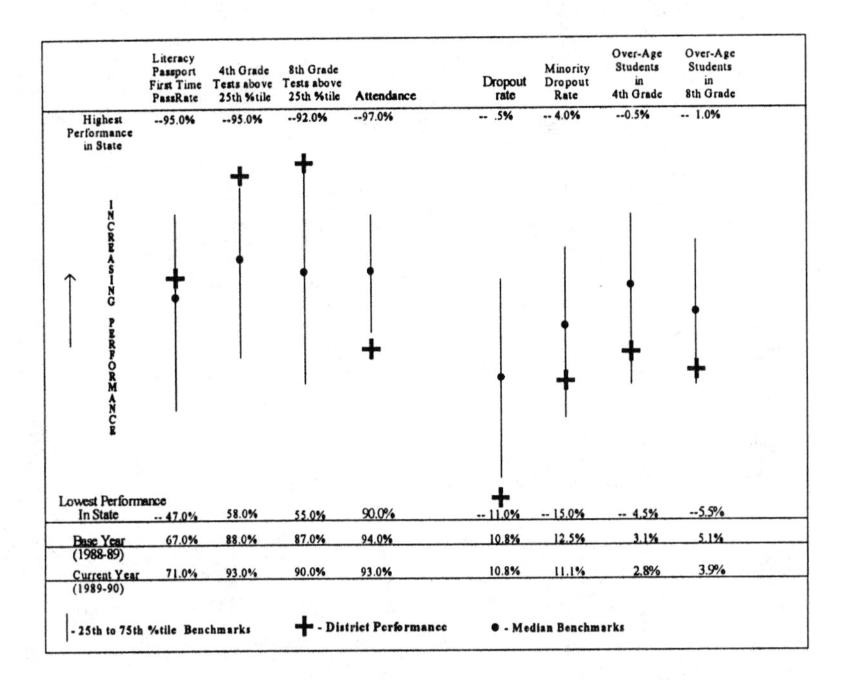

Figure 2.3. School District and Benchmark District Performance
SOURCE: Henry (1993).

this graph were able to retrieve data and make comparisons accurately from this graph and preferred it to a table with the same information (Henry, 1993).

Greater data density is a desirable quality. Tufte (1990) believes that high data density enhances credibility: "High information graphics . . . convey a spirit of quantitative depth and a sense of statistical integrity. Emaciated data-thin designs, in contrast, provoke suspicions—and rightfully so—about the quality of measurement and analysis." (p. 32). But the preference for greater data density needs to be qualified, as in the example presented above. The qualifications come from what we know about information processing, specifically the way long-term and short-term memory function. Chart 2.2 illustrates the process of visual information processing (Kosslyn, 1985; Wilkinson, 1990).

For processing a graphical display, the graph must first be organized into perceptual units that are used in the short-term memory. However, the short-term memory has a very constrained capacity, about four bits

Chart 2.2
Three Levels of Visual Information Processing

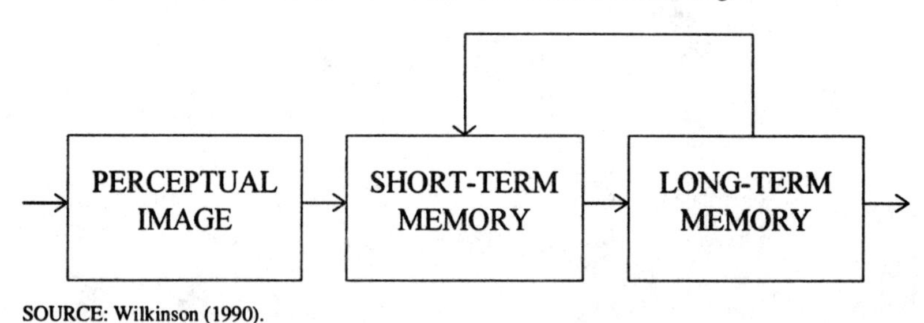

SOURCE: Wilkinson (1990).

of information (Ericsson, Chase, & Faloon, 1980) To process larger amounts of data, the graphical type must be accessible in long-term memory. The symbol systems used to display data—often referred to as graphical icons, graphical types, and graphical design elements—are held in long-term memory and called forth to process the data in a graph (Kosslyn, 1985; Wilkinson, 1990). Bar charts and trend graphs are examples of graphical types. Recall of the relevant graphical type or icon from long-term memory is a necessary step in processing information from the graphical display. As Kosslyn (1985) states, "If one has never seen a display type before, it is a problem to be solved—not a display to be read" (p. 507).

Close inspection of the examples of graphical displays with very high data densities contained in Tufte's (1990) work reveal that most of them are maps or have time as one variable on the *X-Y* plane. These forms are the least abstract and most used displays, indicating that there may be a relationship between the level of abstraction of a particular graphical form and the data density that can be employed. With more abstract, less common designs, it may be necessary to use less dense graphics, which contain more limited units to be stored in short-term memory.

Current theory about processing the data in graphics seems to support the proposition that greater data density may be used more effectively with display types that are stored in long-term memory. Less data should be graphed when the graphical type is less familiar to the audience. The graphs presented in Figure 2.2 are more common than the form used in Figure 2.3 and therefore can be processed more readily. Scatterplots with two or more variables, where time is not one of the variables, are relatively less common in popular publications. As researchers attempt to graph relationships between two variables, such as the relationship between the percentage of students living in poverty and the percentage of students who fail entry tests for the first grade (see Figure 1.6), data densities may initially need to be limited. Concrete examples about what the plot points represent may be necessary. However, after the form becomes more popularized and therefore more commonly present in long-term memory, adding more data may be easily taken in stride. Theory leads to an inference that greater familiarity with abstract graphical forms (education) may lead to increased capacity to use graphics with greater sophistication.

Meaningful data are, in part, defined by the amount of data that can be processed by the audience. Because the familiarity with graphical types and elements varies from person to person and from group to group, limits to the amount of data presented will depend on the audience. This is a soft admonition that stresses the progressive or developmental aspects of graphical display with particular audiences. As more graphical types and elements are stored in long-term memory, more data and sophistication can be comprehended. A researcher's belief that an audience lacks sophistication with graphics is a call for using design techniques that motivate the audience to begin to increase familiarity, not a hard rule for simplistic graphics. The example of the graphic in Figure 2.3 is important in this context. That graph is not simplistic, but the amount of information was limited in view of the lack of familiarity of the audience with the graph type. The next two sections deal with the importance of considering the questions that a graph may be used to answer (graphicacy) on graphical design and the motivation of the audience.

Display Designs

Designing a graphic to present a batch of data is important and difficult. Given the desire to make the display as data dense as reasonable for the audience, we must begin to consider what we wish to encourage the audience to do with the graph. This gives way to a need to understand graphicacy, or what a viewer can draw from a particular graphical display. Wainer (1992) derives from the work of Bertin (1967/1983) three levels of questions that a graphic can be used to answer:

Level 1: What is the level, count, or amount?

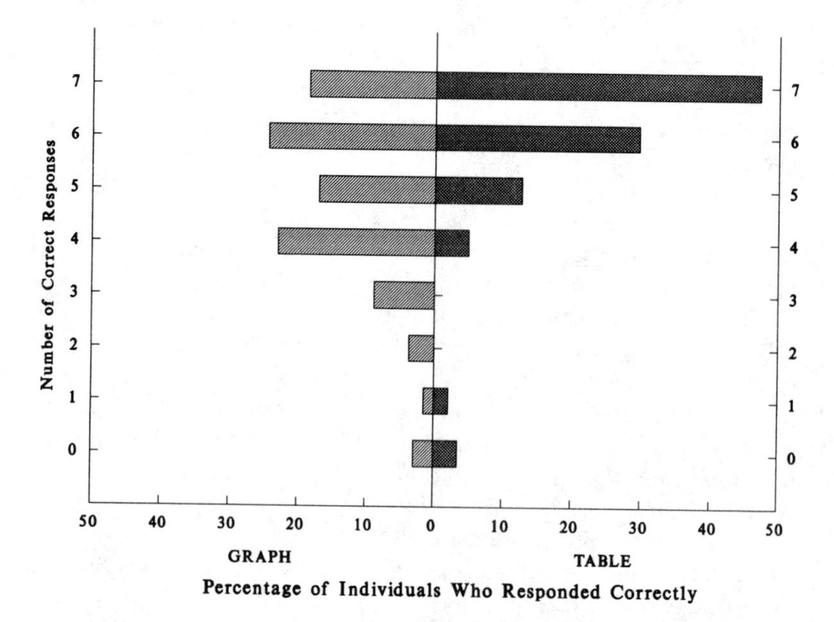

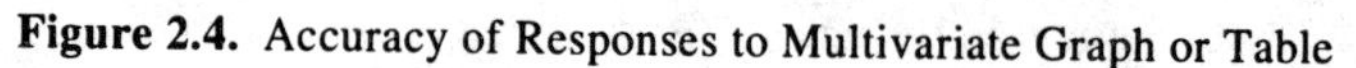
Figure 2.4. Accuracy of Responses to Multivariate Graph or Table

Level 2: What is the norm or trend?

Level 3: How do the trends or patterns compare?

The questions can be answered independently or in combination. For example, in detecting the outliers in Figure 1.6, first we note the norm in the relationship indicated by the line, then we identify those school districts with levels very different from the trend line.

Looking at Figure 2.4, we may pose questions at all three levels. The data presented are the percentage of educators, school board members, and journalists who responded correctly to seven questions that required them to extract information from a graphical display or a table (Henry, 1993). A first-level question is What percentage of those using a graph answered five questions correctly? A second-level question is What was the median number of questions that they got right using the table? A third-level question might be Did they get more correct answers using the table or the graph? Roughly these questions move in complexity from counts (Level 1) to summaries (Level 2) to comparisons (Level 3). Level 1 graphicacy corresponds to using a graph to read numbers instead of a table. Usually, a researcher is better advised to use a table if accuracy in reading numbers is the only purpose for the graph. If a graphical form is chosen for this task, research shows it is only as accurately used as a table when the numbers are displayed on the graph, as in Figure 1.2. A researcher should be able to identify another purpose, and probably more likely, prioritize among a variety of purposes when beginning the design process.

Usually the results of experimental designs of the type that produced the data in Figure 2.4 are summarized by a test of statistical significance, which despite calls to use the *magnitude* of differences (Cook & Campbell, 1979) or confidence intervals (Reichardt & Gollob, 1989), remains the standard for information passed along to the audience. The graph, in contrast, allows the audience to examine the distribution of responses for each type of "treatment" (in this case, types of data displays) and the median or mean for each "treatment group" and to compare the groups at the summary level (comparisons of means) or at a more detailed level (comparison of the distributional patterns). Statistical tests of significance, on the other hand, although taking variation into account, discourage the active consideration, contemplation, and understanding of variation. This observation becomes more alarming when we consider Fisher's (1925) characterization of the object of modern statistics as the "study of variation" (Bennett, 1971-1974).

Later in the book, I will present three possible graphical designs for visually comparing between and within groups: two group histograms (used in Figure 2.4), jittered overlay plots, and multiple box plots. The choice of the type of data display will depend on the data and the graphicacy level at which the audience will deal with the visual display of variation. The latter is determined by audience motivation to retrieve the data and ability to use the graphic, which is discussed in the next section.

Motivation and Interpretation of the Graphical Display

Motivation to gain knowledge from the data and the ability to interpret it accurately are attributes of the audience that must be considered by the researcher in preparing data graphs. Graphical competence is not achieved by preparing an exemplary graphic alone: The graph must be examined by the audience and the data retrieved from it. In revealing the data to viewers, the graph maker must stimulate their interest and facilitate accurate and efficient interpretation of the data.

This task seemingly is made easier by the innate capacity that audience members bring to the tasks: "Graphs work well because humans

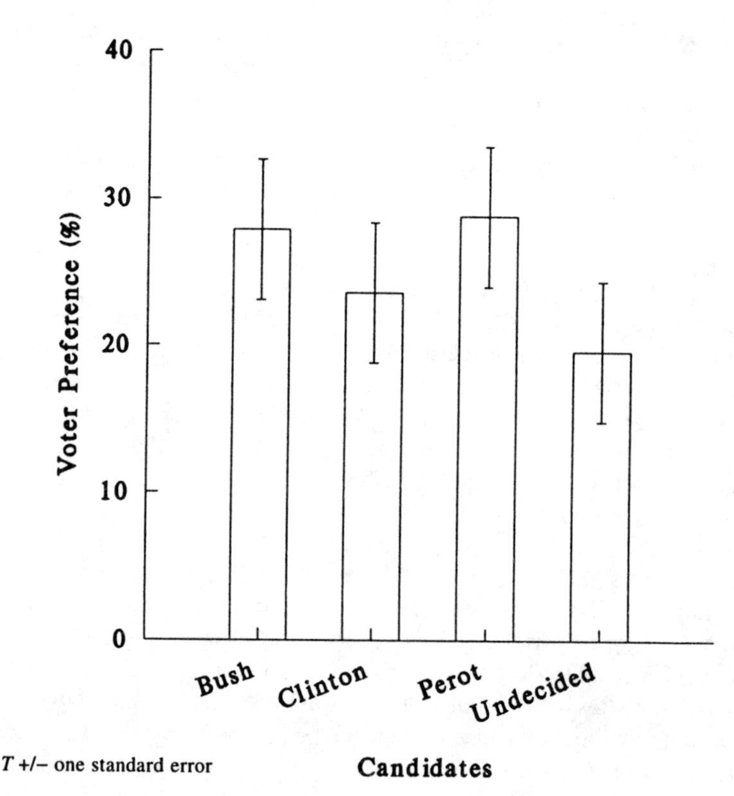

Figure 2.5. Georgia Voter Preferences, Spring 1992

are very good at seeing things" (Wainer, 1992, p. 15). Visual discrimination is basic to us all. Of course, an individual can become more efficient at retrieving data if the form of the visual data display used is familiar to them. That is, developing a facility with graphics should be considered a learning process that has to do with the ease of recall of the particular format from long-term memory. Currently, the facility with ubiquitous bar and pie charts signals that these formats have been imprinted on long-term memory. These are rather simplistic formats and do not necessarily represent the endpoint of graphical sophistication that can be expected from the public. To the contrary, they represent a beginning point from which we can work.

Take, for example, the bar chart shown in Figure 2.5, based on a poll of likely voters in the state of Georgia. Bar length represents the percentage of likely voters who anticipated voting for one of the major presidential candidates in spring 1992. To add more data to the graph, vertical brackets representing error bands were added at the top of each bar. The brackets are drawn +/– one standard error from the sample estimates of voter preferences. (Standard error calculations for samples of various sizes are shown in Henry, 1990.) The brackets emphasize the uncertainties connected with the estimates. The four brackets overlap, indicating there were no significant differences in voter preferences in the spring of 1992. The presidential race was a statistical dead heat at this point in the race. The brackets add this information directly to the graphical displays, important information that is useful for the viewers to know. Yet sampling error rates frequently only make their way into a footnote. Although there are more complicated ways to present the same data (see Cleveland, 1984; Tukey, 1988), the error bars on the bar chart represent a step in the progression toward more sophistication.

The belief in the human capacity to interpret graphical displays needs to be girded up by the moral imperative that we regard the viewers of graphics as being as intelligent as ourselves (Tufte, 1983). They may not be interested in the data. They may not be familiar with a particular graphical form. They may have other demands on their time. But if we get the display right, we must assume that they are intelligent enough to interpret it.

In numerous discussions of graphics with other researchers, I have heard them comment that the idea of more sophisticated graphics is very interesting, but the audience with which they must deal wants only boiled-down, basic information: a graphical sound bite. If we fall into this trap, we do not need to wonder why policymakers and the public have so little regard for the complexity of our research findings. We communicate precious little of the research findings, usually the broadest conclusions that we can comfortably generalize, often cloaked in technical garb. Unless we undertake the process of directly communicating our complex research findings to lay audiences, trusting their intelligence, our work may go underutilized and its value underestimated.

Motivation and the ability to interpret data are important considerations for researchers who wish their research to have an impact. Two principles must be put into play if we are to achieve graphical competence:

1. Audiences are capable of retrieving data from competent graphs and thereby acquiring a greater understanding of the subject.
2. The process of retrieving data from graphs can become more accurate and quicker. It is a learning process: Viewers can learn to use more sophisti-

cated graphics and store their formats in long-term memory for future use if they are interested in understanding the data and provided with a graph appropriate for the task.

These principles underlie the guidelines for competent graphical displays presented in the next five chapters. The chapters are organized by the types of data that applied researchers are likely to collect and, roughly, the most popular statistical techniques that are currently used in each circumstance.

EXERCISES

1. For each of the graphs in Figure 2.2, compute the data density. First sketch out the data matrix that would contain the data for the graph. Compute the bits of information. Compute the area of the graph. Divide the former by the latter.

2. In Figure 2.2, one of the graphs is labeled "Income and Education." The graph summarizes a significant amount of information. Think about the data that was needed for the summary and design (hand draw) three other graphical formats that you would like to see.

3. What is the level of white unemployment in 1970 shown in Figure 2.2? How would you characterize the direction of the trend in black unemployment? How do the levels and trends in black and white unemployment compare?

4. Review the graphs that you selected from popular publications and journals. Were the graphical displays appropriate for their respective audiences? Ask appropriate questions for each level of graphicacy (level, norm, and comparisons) concerning the graphs. What is the answer? What comparisons, if any, do you think the author was encouraging with the graph?

3

Data Summaries: Parts of a Whole

Applied social researchers often use graphs to provide a summary or overview of data that are central to the study at hand. At times, these data come from secondary sources and set the context for the study. The primary purpose of these displays is descriptive. For example, Figure 3.1 shows data that were extracted from an automated data base of an executive branch agency and summarized for use in a legislative oversight commission's report. At other times, the study is the primary source for the data. The graphic may present the dependent variable or describe the population that is the focus for the study. In addition to Figure 3.1, Figures 1.1A, 1.2, and 1.3 present examples of the data graphics described in this chapter.

Graphics that present summary data set the stage for the study. Usually these graphics are not data dense, nor are they at the analytical end of the continuum of purposes for graphical displays. The researchers may use one graph or several. Such displays can serve several purposes for the audience:

1. Orient viewers to the subject or topic
2. Point out a problem or issue central to the study
3. Provide an overview of the group being studied
4. Justify the study's concentration on a part of the topic or a subgroup of the population
5. Allow viewers to generate questions and issues for themselves that the researcher will address in the study
6. Expose a commonly held myth or misconception about the population or topic

In general, this type of summary carries the message of the researcher about some of the facts that will be important for audience members to carry with them as they consider the study. It is a purposeful display and the message contained in it should be clear and readily grasped by the audience. The researchers should use the graphical summary in a

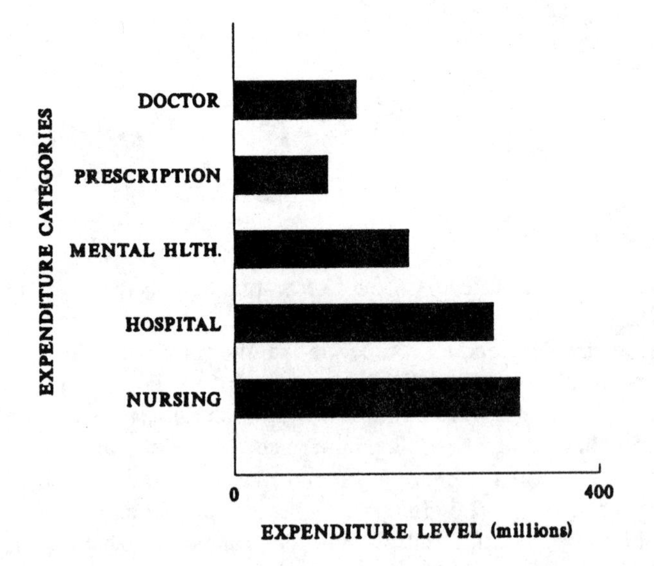

Figure 3.1. Virginia Medicaid Expenditures, 1991

decidedly purposeful manner. In some cases, researchers do not sufficiently think through the purpose of the summary graphics. Rather than a stimulating lead-in to draw the audience into the study, a pie chart with the first categorical breakdown of data that comes to mind, or worse yet, an interesting design that the publications staff comes up with appears in the report. In these cases, another purpose is served—rote decoration. Rather than stamping out a meaningless graph to "break up a page of text," the researchers should ask themselves what audience members will want to know to motivate them to attend to the rest of the study. Sometimes it's the characteristics of the population to establish credibility; sometimes it's a myth-breaking finding that will get them to see the importance of the study that follows.

Graphical forms that are most often used for summary purposes are bar and pie graphs. Pie graphs are limited to showing percentage or proportional distributions of categorical data. Bar graphs can serve up the same data, but they can also be used to convey more complex data, such as responses to a series of items that use a common scale, the values of a single variable over multiple units, or distributional characteristics of a continuous variable (sometimes called histograms) for a batch of data. Examples of these uses of bar graphs appear in Text Graphs 3.1 and 3.2 later in this chapter and in the next chapter, which concentrates on displays for multiple units. Displaying summaries of parts of a whole with bar charts and pie charts as well as some of their variations will be dealt with in this chapter. The stacked or divided bar chart, a variation of the bar chart that is a popular replacement for pie charts because of its similar applications (showing proportions of the whole), will also be included. The stacked bar shows the percentage breakdowns that cumulate to the total (see Text Graph 3.4). Multiple stacked bars can be placed on the same graph to compare different units, and the total length of the bar can remain constant to represent 100% or vary by the total size of the units being examined. All three types of displays are discussed in this chapter.

Obviously, these basic formats are limited in the type of data they can present, especially for pie graphs and divided bar charts. For bar graphs, however, there are a considerable range of uses and there are significant variations in design that should be considered in the development of summaries. A bar chart has an advantage because it is so commonly used—its format is likely to be stored in long-term memory for many members of the audience. Thus, the information processing, as noted in the last chapter, can be more readily accomplished. Despite some of the limitations and perceptual difficulties with bar charts, they are the workhorses of graphical summaries.

Is the Bar Mightier Than the Pie?

Since at least Eells's (1926) article in the *Journal of the American Statistical Association,* statistical graphers have pondered the relative merits of the bar and the pie graph. For at least one task either seems adequate—the breakdown of a variable into distributions by category. Government expenditures by categories (defense, Social Security, debt maintenance, etc.), counts of enrollments by types of high school courses, or percentages of the population in various age groups are all potential data for these forms. Bar graphs have a wider variety of uses than pies, as will be demonstrated later, but in the cases where either is usable, are there data that suggest that one is better than the other?

Putting aesthetic preferences and the desire for variety of presentations aside, two recent studies have empirically investigated this topic. Cleveland and McGill (1984) found that absolute error in judging percentage differences in two slices of a pie was greater than in judging differences in two bars. They also found that differences in size of pie slices were consistently underestimated, but found little bias for bars. Further, large errors were much more likely with pies than with bars. Thus, bars are preferred for purposes of reducing error and bias.

In another study, Simkin and Hastie (1987) found a similar preference for bars, but for a different reason—processing time. Their experiments addressed accuracy and bias and also included time as a criterion variable for preferring one form over another. They found less time required to make judgments about bar length than pie slices. They did not find a difference between the two forms in accuracy. Bars are thus preferred over pie charts in both studies but for different reasons. Figure 3.2, which was adapted from Cleveland and McGill (1984), concretely illustrates the difficulty of discerning the differences between the pie slices (3.2a) and the simple bar (3.2b). First, try to judge the differences in size of Sections C and E using the pie chart. Then, make the same comparison using the horizontal bar chart (3.2b). Both the graphs display the same data, but clearly the bar is more quickly and discernibly different. Thus, for questions that fall into the first level of graphicacy—questions of level, count, or amount—bars are preferable. It is simply quicker and more accurate to read the values from a bar than slices of a pie. Comparing two bars or two slices leads us to the same preference.

Both Cleveland and McGill (1984) and Simkin and Hastie (1987) went further in their studies and considered stacked bar charts. Stacked bar charts are more difficult to make accurate judgments about than either of the other forms. In the other two boxes (3.2c and 3.2d), the divided bar chart is compared to a grouped bar chart for ease and accuracy of determining the order of the proportions. Once again the same data are plotted. Clearly, the horizontal bar (3.2b) and the vertical grouped bars (3.2d) are easier to retrieve data from than their counterparts, the pie and the stacked bar, respectively.

Cleveland and McGill (1984) developed from their experiments and the work of others a hierarchy of perceptual tasks used to communicate quantitative information:

1. Position along a common scale
2. Position along nonaligned scales
3. Length, direction, or angle
4. Area
5. Volume, curvature
6. Shading, color saturation

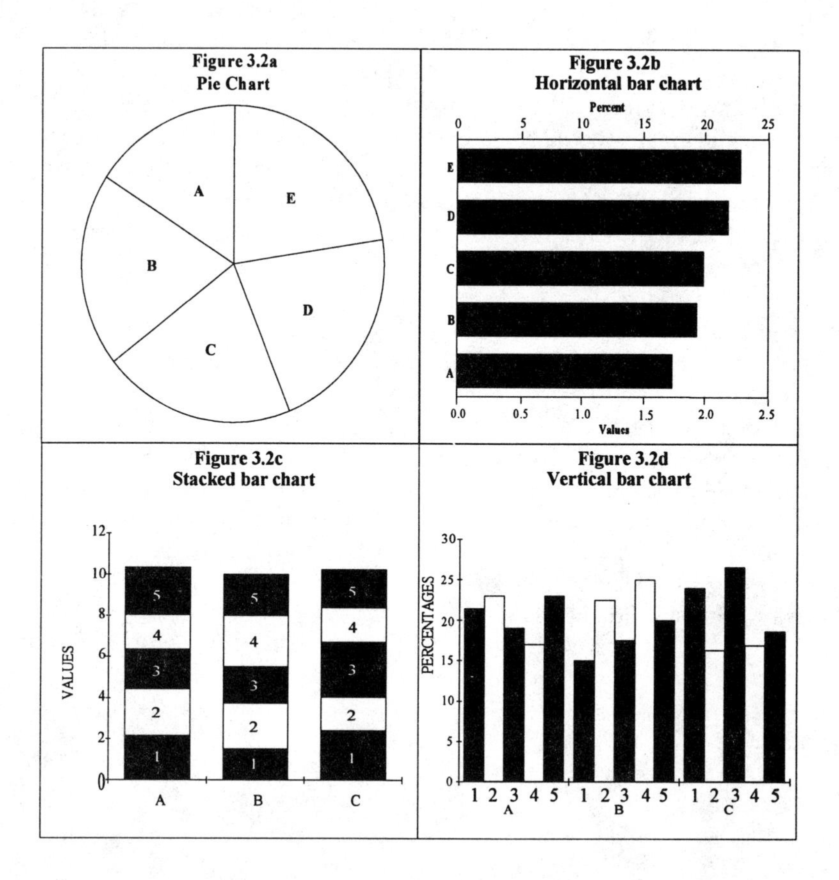

Figure 3.2. Displays of Parts of the Whole
SOURCE: Cleveland and McGill (1984).

Tasks that fall higher on the list are more accurately and more quickly performed. For tasks such as length, direction, or angle that are included on one line, not enough data exist to determine which is preferable for graphic design.

Simple bar graphs fall principally under the first task, but the perceptual task is made more cumbersome by the complication of area perceptions, and in the case of bars with an added 3-D perspective, volume perceptions; pie slice perceptions are for the most part angle tasks, but interference may come from area and length of curve perceptions. Stacked bar charts require information to be retrieved from positions along nonaligned scales. Using this hierarchy, simple bars are preferred to stacked bars, which are preferred to pie charts (although the findings of Simkin and Hastie, 1987, reverse the order of pies and stacked bars). In general, graphs should be designed to require the least difficult task in the perceptual hierarchy. Following this scheme, shading and color hues are the least preferred methods of indicating quantitative differences.

The best method for presenting data graphically would involve only judgments about position along a common scale. Cleveland and McGill (1984; Cleveland, 1984) devised an alternative to pie and bar charts, called the dot chart (illustrated in Figure 3.3), that uses only this perceptional task. The data for Figure 3.3 are the same as for Figure 3.1, but I have displayed a different base, 80 to 100 rather than 0 to 400, to highlight an advantage of the dot chart. Because the dot chart conveys the quantitative information through the position of the dot along a common scale, it does not mix length and position perceptions the way the bar chart does. Therefore, the base for the dot chart does not need to be zero. In practice, the convention of a zero base for bar charts is often violated, but this can send mixed messages about the data, often exaggerating differences.

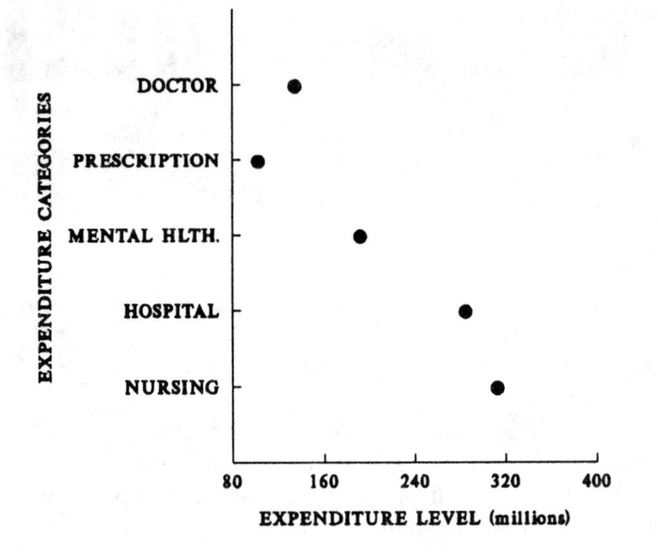

Figure 3.3. Virginia Medicaid Expenditures, 1991

Unfortunately, Cleveland and McGill (1984; Cleveland, 1984) do not present empirical evidence of the dot chart's superiority. Until this is done, it seems advisable to use the bar chart whenever possible. It has empirical support and is likely to reside in the long-term memory of many in the audience for the graph. For those who persist in using pie charts, in the next section I will briefly suggest some design elements that should be avoided.

Pie Charts and Other Distortion-Prone Formats

Simple, unadorned pie charts, such as the one shown at the top of Figure 3.4, have been shown to present more perceptual difficulties than alternative graphical formats. For this reason they are shunned by many graphics experts. They do what they were designed to do; as Bertin (1967/1983) notes, "the essential point of the information—the relationship to the whole—is depicted" as long as no external comparisons are needed and the categories are few in number (p. 200). But as used in practice, pies often present design features that further impede audience comprehension. Figure 3.4 illustrates seven common problems that cause data to be distorted: fake 3-D perspective, too many categories, moire effects, imperceptible shading contrasts, ambiguous labels, legends, and arbitrary pullouts. Rather than enabling a graphic to reveal data, these elements can distort comprehension.

A fake 3-D perspective in a pie chart— or other graph—is usually seen as more artistic or adding a more professional "look." Schmid and Schmid (1979) comment, "Correct projection technique and skilled artistry, combined with a knowledge of constructing statistical charts, can produce an authentic and esthetically appealing pie chart" (p. 293). Some software programs add the perspective as the default option, it has become so popular. However, careful examination of the second panel of Figure 3.4, which was adapted from Schmid and Schmid (1979, p. 260), shows the problem of distortion in using projection techniques for achieving the illusion of a third dimension. Military spending appears to be much more than 60% of the chart. Because of the distortions of the projected, pseudo three-dimensional image, to my eye, the slice for 7% appears to be less than one half the size of the 15% slice.

Tufte (1983) has labeled phenomena of this type as the "lie factor," for which he provides a measure:

$$\text{Lie factor} = \text{size of effect in graphic}/\text{size of effect in data}$$

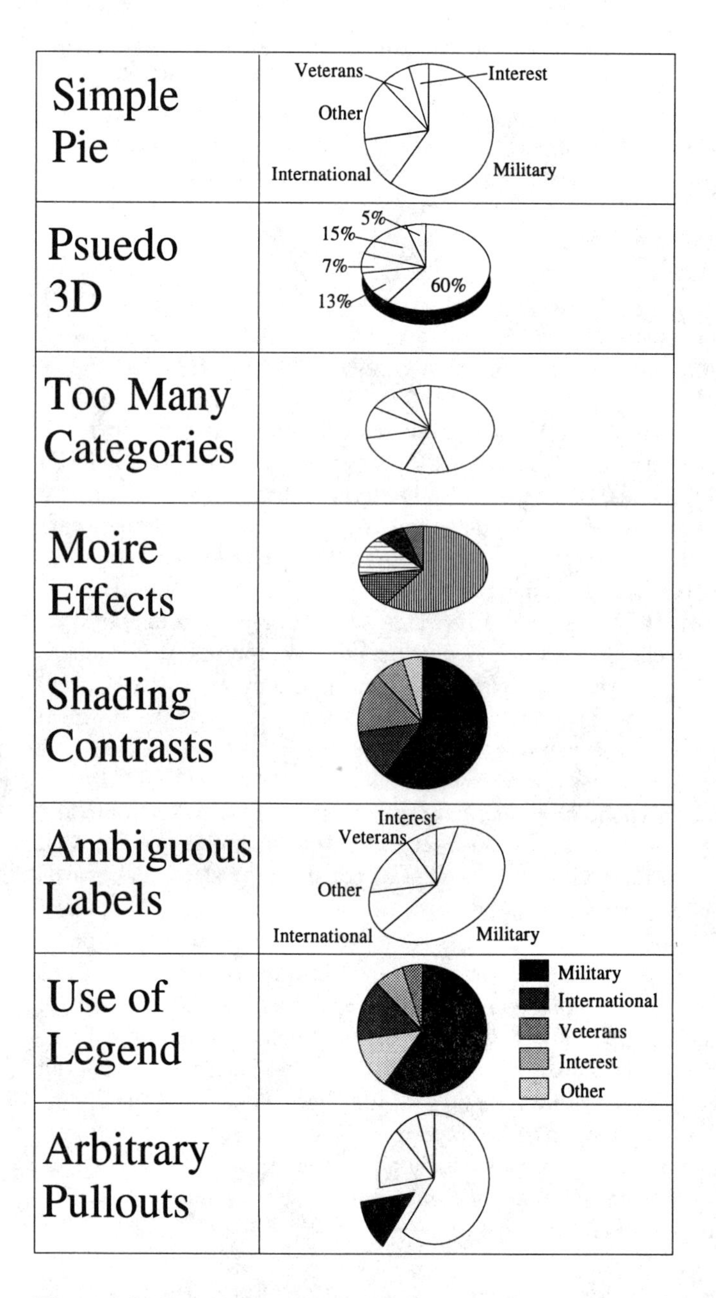

Figure 3.4. Common Problems With Pie Charts

Figure 3.4 is not as much wrong as it is ambiguous. Cleveland and McGill (1984) conclude that pie charts primarily involve the task of judging angles, but area and length of the circumference become other potential ways for decoding the data. With three-dimensional perspective, area as a way of decoding the data differences further confounds the perceptional tasks. The graph is accurate if viewed from the first perceptual perspective, but distorts the data if viewed from the others.

Because the researcher cannot control the perspective that an audience member will take, it is best to use a graphic that provides less ambiguity. In many ways this is similar to editing items for a questionnaire to remove ambiguity so that the meaning is the same for all respondents. Researchers should be as concerned with ambiguous graphics as they are with ambiguous questionnaire items—both distort the data. Tufte's (1983) solution to the graphical problem is straightforward: "The number of information carrying (variables) dimensions depicted should not exceed the number of dimensions in the data" (p. 71). This suggestion leads us back to the bar chart, or as Tufte (1983) might suggest, a line graph for these data (p. 101). In a line graph, a line—which is perceived as one dimensional—simply replaces the two-dimensional bar, which can be slightly ambiguous due to the area and length perceptional possibilities.

Without question, the 3-D effect for pie charts should be eschewed by researchers. In similar fashion, pictographs that are ambiguous and distort the effect size in the data should also be avoided. Pictographs use pictures such as soldiers or lightbulbs in place of bars or lines to represent quantities. It is important to recognize the potential for distortion in these pictographs. Tufte (1983) presents a number of examples of pictographs that distort the effect size (pp. 57-71). Schmid and Schmid (1979) provide three versions of a pictograph that show the size differences depending on whether height, area, or volume is used to convey the data (pp. 221-223). Because pictographs are found much more often in the popular press than applied research reports, I will not deal with them further in this book. I will, however, point out that the practice of decorating a bar chart with a picture on the side of the graph, as is sometimes done in *USA Today,* is much more innocuous than using the picture to convey the data, as is often done in the same publication. In general, any graphical representation that has a strong potential to distort data should not be used.

The other problems with pie charts illustrated in Figure 3.4 can be remedied. Six categories are the upper limit for pie charts (Bertin, 1967/1983, p. 199). If there are more categories some should be combined for a total of no more than six. A moire effect is the illusion of curved or wavy lines that stems from the use of hatching and cross-hatching. This headache-provoking effect can be eliminated by using shadings of gray, from black to white, in place of cross-hatching. By arraying the shadings to provide maximal contrast from one slice to the next, the imperceptible differences between shadings can be reduced. Ambiguous labels can be eliminated by using labels on the side that are attached by lines to their respective slices. Labels of this type eliminate the need for legends, which take more time to process and can lead to ambiguities because of the difficulties in perceiving differences in shades of grey. Pullout slices are often used to emphasize a specific category that the researchers deem to be important. However, graphical software programs sometimes default to pulling out the largest slice, no matter which one the researchers wish to emphasize. Exploding slices should only be used at the prerogative of the authors and then only when it makes an essential point in the study. These suggestions would lead to a pie chart similar to the one that appears in Figure 3.5 and displays the same data as in Figure 3.1. However, in the pie chart, the level of expenditures is lost and only the relational aspects of the expenditures are displayed.

Bar Graphs

Bar graphs are the most highly recommended format for summary data. Data from the previously mentioned studies support this, although graphical experts have suggested variations that from a logical standpoint seem to be improvements (for example, dot charts and line charts). However, logic and perception seem to depart company in some cases. Psychologists interested in the differences between actual size differences and perceived size differences have empirically estimated functional relationships between the two that are usually discussed as Weber's law or Steven's power law.

Typically, actual size increases are underestimated, which causes the differences to be distorted. "The perceived area is usually proportional to the actual area raised to an exponent of about .8" (Kosslyn, 1985, p. 504; see also Cleveland, Harris, & McGill, 1983; Cleveland & McGill, 1984; Tufte, 1983). In a review of many experiments, Baird (1970) showed that the negative bias in distortion is effectively 0 for lines, grows for area (.8), and is even larger for volume. The work of Cleveland and McGill reinforces the negative bias that distorts the perceived size of the effect in pie charts; they find no such bias in the perception of bar

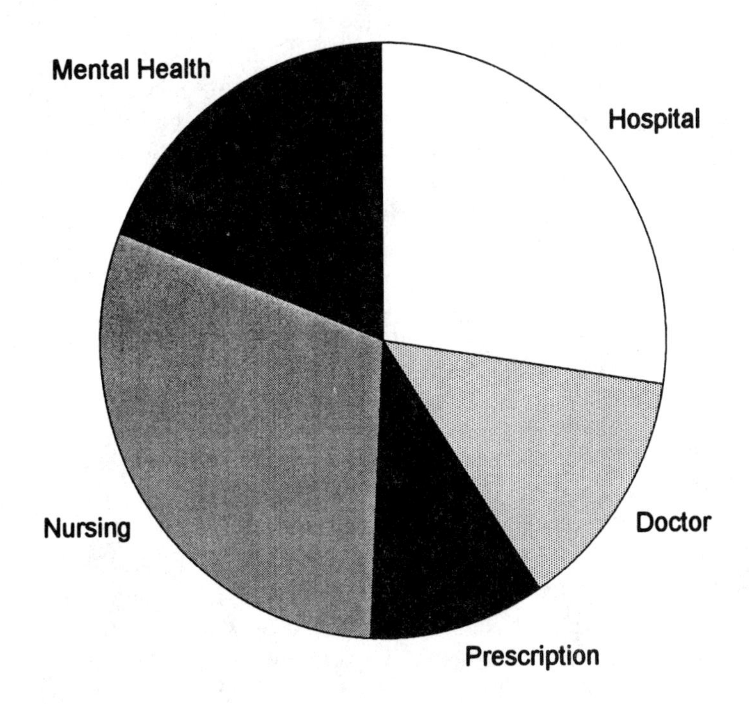

Figure 3.5. Virginia Medicaid Expenditures, 1991

charts. Other graphical formats have yet to be tested, and therefore their likely perceptional distortions are unknown. In this case, we can follow the old adage, "If it ain't broke, don't fix it." But we should begin to establish an empirical base for these alternative formats as they may be useful innovations.

Bar graphs do have a number of design variations that should concern us. Orientation, grid lines, axes, tick marks on the axes, fill, and order of the bars are considerations that shouldn't be left to software default settings.

Orientation. Orientation refers to whether the bars run vertically or horizontally. Figure 3.1 has been transposed into a vertical bar chart in Figure 3.6a and repeated as a horizontal chart with grid lines in 3.6b. Little research has focused on the question of orientation, but several studies have shown inconsistent results in judging lengths based on the alternate orientations (Behrens, Stock, & Sedgwick, 1990; Stock & Behrens,

1991). John Tukey, who fueled a renewed interest in graphical analysis with his book *Exploratory Data Analysis* and the considerable volume of graphical work that followed, believes that a horizontal line is a powerful base from which people make judgments (Tukey, 1988). Although this advice is offered in another context (the question of smoothing), it appears from examples in the text that it could be a nod toward vertical bars, where viewers compare length by visually drawing a horizontal line or using the horizontal grid lines. However, it is interesting to note that Cleveland and McGill (1984) tend to use horizontally oriented graphs to show their results and to illustrate "good" practice. One major advantage of horizontal bar charts is the ease of reading labels set on the left side of the bars, rather than labels below—often abbreviated or set at an acute angle. Although the prevailing convention for one type of bar charts, frequency distributions, favors the vertical orientation, it may be a good practice to use both in a report or paper for variety, but stick to the one you find most appealing in any section.

Grid Lines. Grid lines are highly controversial among graphical designers. Figure 3.6b shows the same data with grid lines. Most standards call for grid patterns, but offer little guidance about their number or spacing. Tufte (1983) strongly suggests the minimalist approach. For bar graphs, it should be obvious that grid lines running parallel to the bars are unnecessary. Remaining grid lines should be printed as lightly as possible. With a few bars, many software packages allow the percentages to be written in the bar or just beyond its reach. Level 1 graphicacy becomes a matter of reading the number, making interpolation based on grid lines unnecessary. Research has shown that with the numbers provided on the graph, readers use graphs as accurately as tables, but without the numbers, tables are more accurately used (Jarvenpaa & Dickson, 1988).

Axes and Tick Marks. Axes provide a frame for the bars and to some eyes give the appearance of "completeness" to a graph. To my eye, they act as arbitrary caps on the percentages (or values of the variables in other graphs). Figure 3.7a shows the data from previous graphs with four axes. Tick marks are commonly included on the axes, as is the case in Figure 3.7a. Many familiar with using graphs will find the absence of axes lines and tick marks disconcerting, though they are not needed as an aid to interpolation when the values are printed on the graph.

3.6a

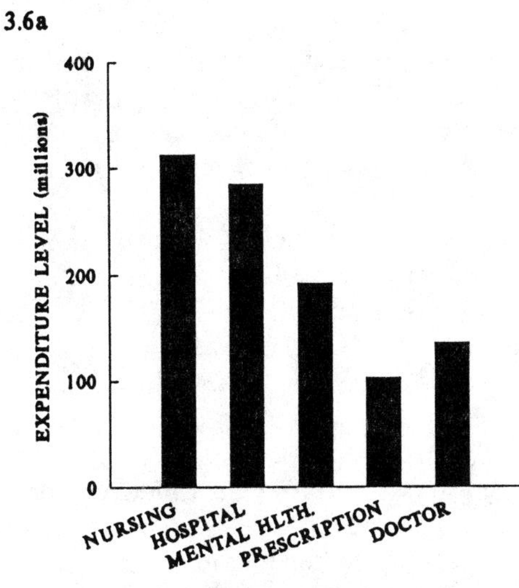

3.6b

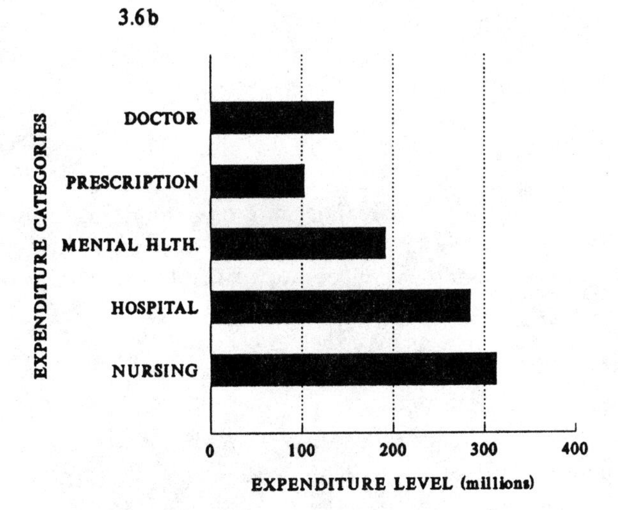

Figure 3.6. Virginia Medicaid Expenditures, 1991

Fill. The final option for simple bar charts I present is fill. Two points should be made here. If filling in bars encourages the audience to consider differences in the area of a bar rather than the length, fill should be avoided. Without filling, as shown in Figure 3.7b, the graph appears anemic, less dramatic. To gain the advantage of fill without encouraging area comparisons, I prefer thin, filled bars as shown in Figure 3.1.

Order. Figure 3.7b also illustrates one other option: the order of the bars. In this case the expenditure categories are presented in ascending order, from smallest to largest. Bertin (1967/1983) suggests ordering to imprint a pattern in the mind of the viewer. By reading the categories from top to bottom, the viewer has obtained ordinal data about the categories. Nursing expenditures are the highest and prescriptions are the lowest. The magnitude of the differences is then quickly grasped from the length of the bars. The choice of ascending or descending order depends on the data and the point the analyst intends to illustrate with the graph. In this case ascending order provides the most proximate interpolation of the largest expenditure category, nursing, and this category becomes a base to judge the relative size of the other categories as the viewer scans down to the axis to judge the level of expenditure in a particular category. For data displaying multiple units, descending order has the virtue of putting the largest unit on top and emphasizing the large units, as we tend to read from top to bottom.

Beyond the Basic Bar

The basic bar chart, presenting parts of a whole, is an extremely useful tool for data summaries. Researchers can take advantage of the format of the basic bar to introduce more data into a summary or at other points in the study presentation. Five uses are presented in Text Graphs 3.1 to 3.5 that researchers may wish to consider:

1. Frequency Distributions
2. Multivariate items on the same scale: selected values
3. Multivariate items on the same scale: all values
4. Parts of the whole for three groups: paired versus grouped
5. Parts of the whole for two groups: paired and grouped

In each text graph, the description of the use, a common situation or situations where the use might be applicable, and an example graph are

3.7a.

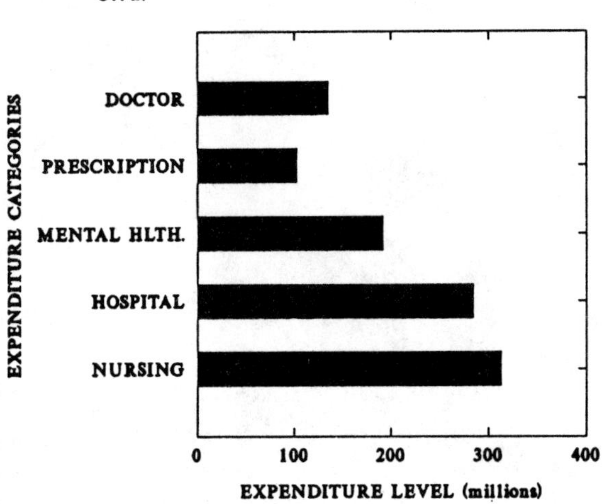

3.7b.

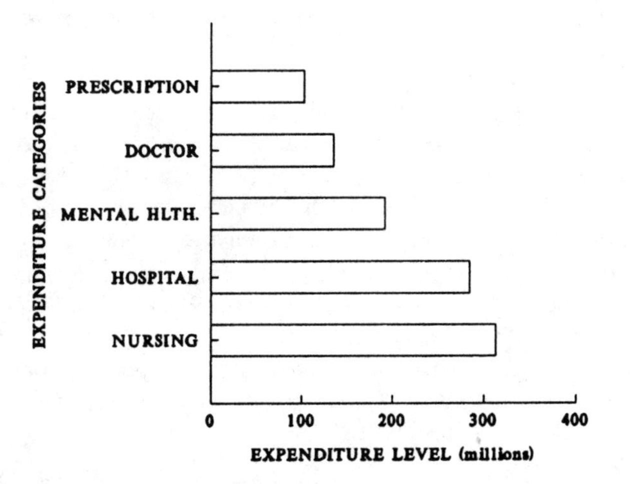

Figure 3.7. Virginia Medicaid Expenditures, 1991

Text Graph 3.1
Frequency Distribution

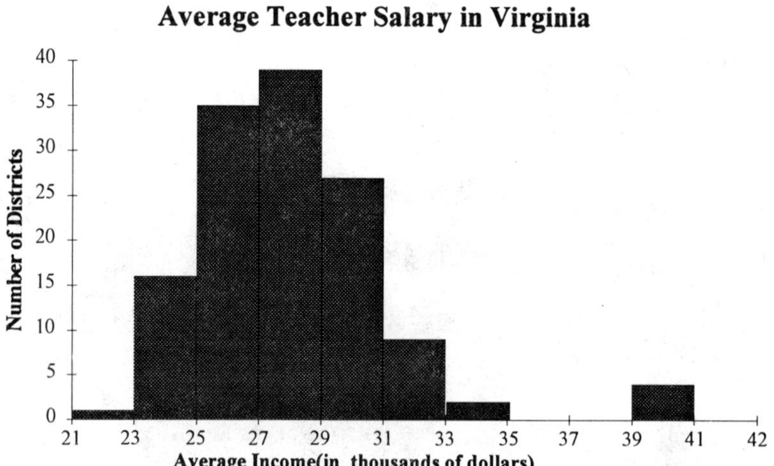

The data presented are average teacher salaries for the 133 school districts in Virginia. Any continuous variable can be categorized by the number or percentage of units that fall into equal size intervals. The graph designer can choose the width of the intervals, which becomes the width of the bar; in this case $2,000 intervals were chosen. Also, the graph designer can choose whether to plot the number or percentage of districts in the interval.

Bertin (1967/1983) makes a useful point about using this graphical format—it is easy to recognize null values, that is, there are no districts with average teacher salaries between $35,000 and $37,000.

presented. The five examples are not exhaustive of the uses of the bar chart for data summaries, but are meant to encourage more thought and creativity.

Text Graph 3.1 presents teachers' salary distribution data for the school districts in Virginia. The same data are presented in Figure 1.3, but there dots are used instead of bars. For frequency distributions such as this, it is customary to orient the bars vertically and remove the spaces between bars that are typical of other bar charts. Most statistical software and spreadsheets can generate frequency distributions. To compose the graph, the continuous variable must be divided into equal size intervals. Most software programs allow the analyst to choose the interval size, but default to an arbitrary number of intervals if no choice is made. Text Graph 3.1 displays $2,000 intervals. The example displays number of districts on the vertical axis. Alternatively, the percentage of units can be shown or both can be shown by displaying two vertical axes, one with each scale. (See Figure 3.2b for an example of using two axes, one representing the number of units and the other the percentage of units).

Text Graph 3.2

Multivariate Items: Same Scale Using Selected Values

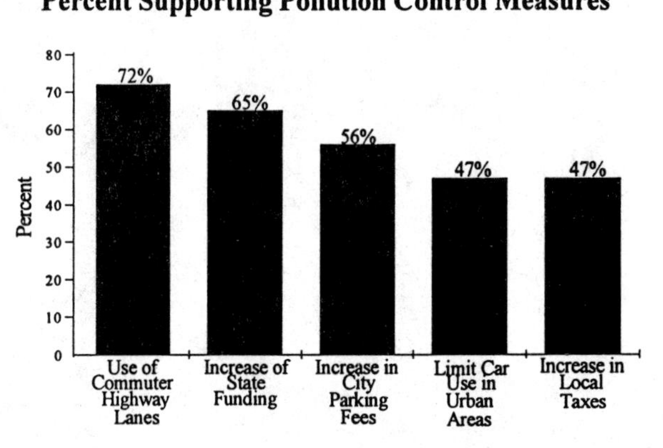

The data presented here are percentages of Georgians who agree with five pollution control measures. Often in survey data, a number of items use the same response categories. One category can be used, or two or more categories can be combined. In this case, agree and strongly agree were combined.

Text Graph 3.2 illustrates the use of a bar graph for showing multiple variables that are measured on the same scale. In this case the scale is a 5-point Likert-type scale. The percentages of respondents who agreed or strongly agreed with each of the five items is displayed. The graph visually emphasizes the level of positive responses to each of the items, or in other words, encourages between-variable comparisons. The percentage estimates are written in above each bar to allow the viewer to quickly complete the task of finding the support for each measure (Level 1 graphicacy). The variables are ordered such that in reading from left to right, the level of support decreases. This conveys information in the order of the variables and makes the perceptual tasks of determining the "norm" and comparing the levels easier.

Text Graph 3.3 illustrates the effective use of the stacked bar, or divided bar graph. Both terms for this type of graph can be found in the literature: divided bar is more accurate in that the bars, as in this graph, are not literally stacked, but stacked bars are more commonly referred to in graphics software. Stacked bars are best used when the data are to be divided into three categories. Where two categories are reported, only one category needs to be displayed, such as the support category in Text Graph 3.2; the percent not supporting the measure can be inferred. Displaying three categories of responses allows two categories

Text Graph 3.3
Multivariate Items: Same Scale Using All Values

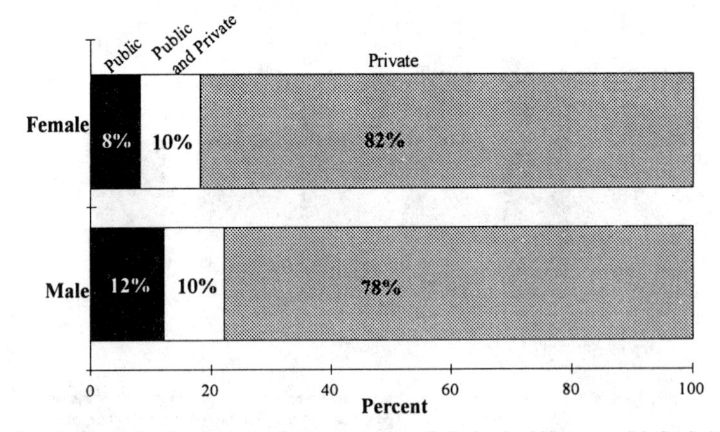

The data for this stacked bar chart came from a survey that asked who should be responsible for decisions about abortion. Men are more likely to believe it is a public decision and should be regulated. Divided bars are useful when the information to be displayed is not easily divided into two categories, as was the case with Text Graph 3.1.

to be aligned on a common axis, in this case the percentage who believe the decision should be a governmental or public one and those who believe it is a strictly private one. One purpose of this text graph could be to undermine the hypothesis that women and men feel differently about the locus of abortion decisions.

Text Graph 3.4 shows two versions of displaying identical data, principally for comparing between groups. In the first panel, the bars are grouped together by the variable (church attendance) that is expected to be related to the response percentages (responsibility for abortion decision). This grouping emphasizes the comparison within the three levels of church attendance. In the lower panel, the response categories are put together, emphasizing the differences in responses between the three groups. A relationship between amount of church attendance and locus of the decision for abortion seems evident. Grouping structures the way comparisons are most likely to be made: The more proximate the location, the more likely the perceived grouping and comparison. In this example, the lower panel encourages the viewer to conclude that those who attend church weekly are least likely to view the abortion decision as strictly private and most likely to view a role

Text Graph 3.4

Multivariate Items: Parts of the Whole for Groups

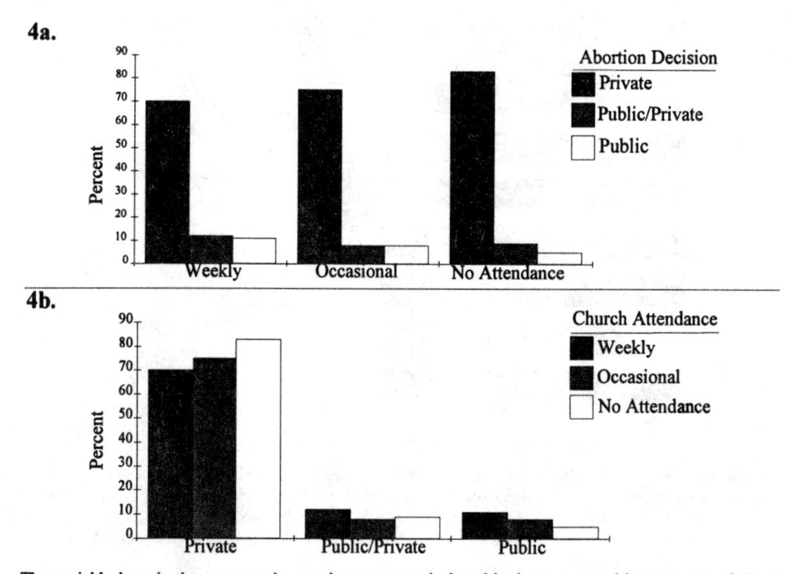

The variable here is the same as the previous text graph, but this time presented by category of church attendance. The grouping of the bars by church attendance encourages comparisons within attendance groupings. The grouping by the response category makes the comparison between attendance categories easier. The choice of grouping should be based on the researcher's judgment about the priority for comparison.

for government in deciding the issue. Gestalt psychologists of the 1920s first established the proximity grouping as one of the Gestalt Laws of Organization. It is most useful for guiding the organization of data into graphical displays. Either of the designs in Text Graph 3.4 will allow the consumers to make the comparisons, but the grouping will lead to one being made most readily. In the top panel, one quickly sees that a strong majority of all three groups believe the abortion decision to be a private one. However, if the ability to explain abortion policy decisions is the objective of the graph, the lower panel makes the appropriate comparisons proximate. The proximity rule is one that I will return to in Chapter 4, where I will show the price paid for violating it for substantive reasons.

The final text graph shows an attempt "to have our cake and eat it too" by encouraging both types of comparisons, within group and between group, for bigrouped data. The graph aligns two sets of horizontal bar graphs and lines them up along a vertical spine. The spine

Text Graph 3.5

Parts of Whole for Two Groups: Paired and Grouped

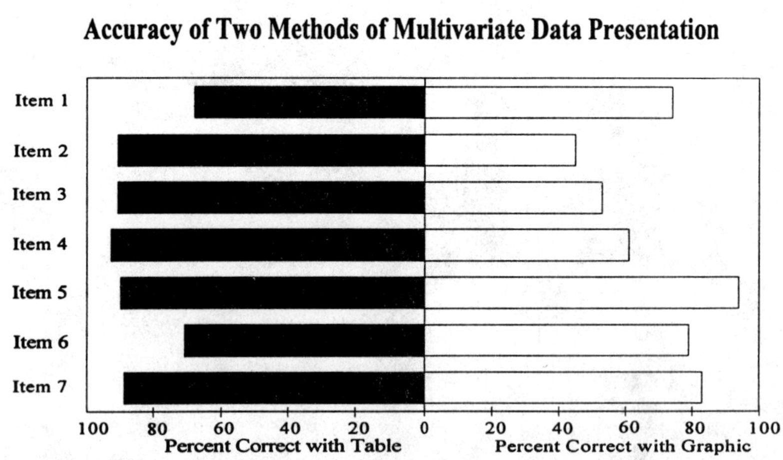

The data presented are from an experiment in graphical perception. The black bars show the percent of individuals who answered correctly using a table. The unfilled bars show the percentage of individuals who answered correctly using a graph. The graph makes item analysis for the two groups easy. Items 2, 3, and 4 were more likely to be answered correctly using a table. Item 1 appears to be difficult for both groups.

separates the group that used a graphical display from the group that used a table in an experiment involving the accuracy of information retrieval. The display quickly shows that Items 2, 3, and 4 were the least accurately responded to by the group using the graphic. Each of those items was responded to correctly by a larger percentage of those using the table. For the other items (1, 5, 6, and 7), little difference appears between the group with the graphic and the group with the table. Both within-group and between-group comparisons are proximate in this design. This design is likely to work best when the audience is reasonably practiced with graphics and the researcher wishes to encourage careful viewing of the data.

SUMMARY

This chapter contains important considerations for data summaries and some ideas for "best practice" using graphical displays. The most common graphical formats that are used for data summaries are bar charts, pie charts, and stacked bar charts. Research clearly indicates that bars are to be preferred to pie charts or stacked bar charts. Further, pie charts are often made more difficult to interpret because of design problems, the most significant of which is fake 3-D perspective. The fake 3-D perspective causes graphical distortion, which is measured by the "lie factor," and results in giving a false impression of the effect size in the data from the graphic. Other sources of ambiguities and visual distortions, such as inadequate labeling and pictographs, are illustrated in the chapter.

Because of the strong support in research findings for the use of the bar chart, the options for its use are presented and illustrated. Although there is no empirical basis for choosing among the options, the researcher will want to be aware of the options for preparing what appear to be the most effective data summaries. The last part of the chapter focuses on the use of bar charts for summaries that go beyond displays of parts of a whole to depictions of parts of the whole for two groups. Often, however, researchers need to find displays that can present data from more than two groups. Multiple treatment groups, multiple program units, several jurisdictions, or regional summaries are examples. Graphical displays for these types of data are discussed in the next chapter.

EXERCISES

The following table contains the actual expenditures of the federal government for fiscal year 1992 (U.S. GAO, 1993):

Budget Category	*Expenditure (\$billions)*
Discretionary	
National defense	301.3
International	19.2
Domestic	213.8
Mandatory	
Social Security	285.1
Medicaid	67.8
Medicare	116.2
Unemployment compensation	37.0
Other	181.2
Net Interest	199.4
Total Outlays	1,421.0

Enter this data into your spreadsheet or graphics software program. Use the full labels.

1. Using the nine categories that have expenditures listed in the table and the default settings of the program, generate a bar chart, pie chart, and stacked bar chart. Analyze the graphs for the suggestions given in this chapter.

2. Change the orientation on the bar chart. Which way are the labels easier to read? Change the fill setting. Order the bars from smallest to largest expenditures.

3. Remove the 3-D effect from the pie chart if it is included in the default or add it, if not. NOTE: Some software will not include a 3-D option. Experiment with different methods of labeling—shading the slices and hatching the slices with legends; labeling outside the pie near the appropriate slice; and labeling with a line attached to the appropriate slice.

4. Using the three major breakdowns, discretionary, mandatory, and net interest, design three separate pie charts, three separate sets of bar charts, and three stacked bars. What is lost in the pie charts? What are the advantages of the bar charts and the stacked bars? Experiment with the bar charts until you have a graph you believe effectively communicates the data.

5. You are beginning a report on the effect of net interest payments on federal expenditures. Design and execute a graph to provide a context in the introduction.

4

Displaying Multiple Units

Applied social researchers often have data on cases that they wish to be able to compare case by case. For example, educational researchers now collect data on student achievement and attainment at the school and school district level. To understand how patterns of outcome variables differ from one school to the next, it is important to compare them. Graphical displays of these data could reveal patterns in the data as well as communicate performance levels to a wide variety of people interested in schools and education. The literature is replete with other examples. In experimental settings, comparing multiple treatment groups across several input characteristics or multiple measures of expected outcomes is often desirable. Graphical comparisons can express patterns and substantive differences in ways that complement and enhance traditional statistical analyses.

Applied researchers generally analyze the data and report the results of the statistical analyses as estimates, sometimes with confidence intervals and statistical tests of significance. These analyses are summaries, and as such, they provide very little data about specific cases. We may know the average of all cases or the percentage of variance explained by an independent variable, but we do not find out if a particular case fits the pattern or if it is relatively high or low or in the middle of the pack. Usually, when the researchers deem it important to provide the results for individual cases, they do so by recording them in tables. Before turning to graphical approaches, the more conventional approaches should be considered.

Data Tables. We can provide results with tables that display the data for each case or unit to audiences for applied research and let them ferret out the conclusions through their own analysis. Often, applied research and evaluation reports (and reports of social statistics) are filled with pages of tables. Although the intention of providing the consumers with the "actual data" is admirable, their access to the information is limited by the format. To use an example from the field of education, I can imagine that the teachers and principals in a school might become more interested in school performance data if they could view the level of performance of their school, get a sense of norms or patterns in the data, and compare their school to other schools; that is, if they could easily extract information corresponding to the three levels of graphicacy described in Chapter 2.

However, if motivated consumers such as these exhaust their time, energy, and capacity scouring a table without understanding the relationships or performance of their school, or come to unwarranted or inaccurate conclusions, we have not adequately supported their desire to obtain useful information. We may take some comfort that at least we gave accurate data as we tell ourselves that we are not responsible for how they use it.

Data tables are excellent methods for storing data, but difficult media for communicating and analyzing data. Wainer (1992, p. 21) has provided useful suggestions for improving tables:

1. Order the columns and rows in a way that makes sense, such as by placing cases and variables likely to be of most interest first or by natural sequences.
2. Round the data as much as possible, as two digits are about all humans comprehend and all our data will justify.
3. Include summaries and averages as a means for making comparisons.

Table 4.1 illustrates these suggestions using some school district performance data from 14 school districts in Virginia. The group was selected by a benchmark selection technique that uses five variables to select the districts that are most similar to a "seed" district, in this case Fairfax County, the district highlighted by gray shading (Henry et al., 1992). In Table 4.1, the 14 districts have been ordered based on their poverty level, as measured by the percentage of students eligible for free lunch. Fairfax County is sixth in poverty, slightly lower than the median, with 13%. Poverty being related to school performance sets up a rough indicator of expectations for judging the performance of these districts, which have already been determined to be similar. Looking at the median for each variable that appears at the bottom of the table, Fairfax County schools can be quickly seen to be above the median for all five performance indicators, and in fact are in the top four on every indicator as shown by the rankings in parentheses.

The indicators that measure participation in foreign language courses are grouped together so that a comparison can be quickly made between overall participation and participation of minority students. Although Fairfax County ranks well on both indicators, the proximity allows one to note that Fairfax County has the lowest difference in the two percentages, indicating a substantial degree of "equity." The suggestions made by Wainer (1992), along with the inclusion of rankings, the use of a predictor variable to order the results, and highlighting the district of interest can improve the audience's ability to access information contained in a table, but they do not make it easy to perform analysis with tables, especially if there are more cases than in this example.

Summary Statistics. A second option is to analyze the data and provide summary conclusions to the consumers. The results of a statistical test of a hypothesis, the interpretation of the direction of a relationship using a regression coefficient, or an estimate of the effect size or confidence interval (Reichardt & Gollob, 1989) are methods that summarize data and provide some guidance about the significance and magnitude of relationships.

But the technical artifacts of these relationships do not give consumers as much information as they may need. In addition, the consumer of the information becomes a captive of the researcher's analysis and rationale. More sophisticated analysis distances the user of the study further from the data itself. Yet without a more sophisticated analysis, the consumer can be misled about program success. For example, looking at measures of students' achievement without taking the level of poverty among their families into account can confuse the advantages of privilege with the impact of schooling.

Finally, these summary analyses do not provide information on specific cases or program sites. For example, analysis of residuals (actual minus predicted outcome measure scores) gives attention to performance of an individual unit after controlling for other variables. However, residual analysis requires some technical sophistication to understand and may engender undue precision in the results, especially when the percentage of the variance explained is low. Accuracy of the predicted value is dependent on the explanatory power of the regression model. The predicted value is, after all, a point estimate and should be surrounded by a confidence interval.

Although both of these approaches, data tables and summary statistics, are essential for the research process, for very different reasons neither is completely satisfactory for promoting the probing of the data by interested audiences. This seems particularly true for people interested in comparing a specific case to other cases to better understand the particular case. One can imagine that a well-designed graphical display could provide those motivated teachers and principals relevant

Table 4.1
Student Performance Data for Fairfax County School District and Its Benchmark Districts

	Students Eligible for Free Lunch	*Absent Less Than 10 Days (6th-8th Grade) Attendance*		*Literacy Passport 6th Grade Pass Rate*		*Taking Foreign Language by 8th Grade*		*Minority Students Taking Foreign Language by 8th Grade*		*8th Grade Standardized Test Scores Above the 75th Percentile*	
1. POQUOSON	4	79	(5)	68	(12)	21	(13)	*	*	53	(1)
2. CHESTERFIELD	10	75	(7)	76	(5)	36	(11)	24	(9)	38	(10)
3. ROANOKE CO.	10	89	(1)	70	(8)	44	(6)	23	(10)	40	(8)
4. FALLS CHURCH	11	79	(5)	80	(2)	58	(1)	35	(6)	53	(1)
5. STAFFORD	11	66	(13)	72	(7)	30	(12)	16	(12)	38	(10)
6. FAIRFAX CO.	12	81	(2)	79	(3)	46	(4)	42	(2)	53	(1)
7. PRINCE WILLIAM	12	67	(11)	69	(10)	40	(9)	29	(7)	40	(8)
8. MANASSAS	14	53	(14)	83	(1)	20	(14)	9	(13)	44	(5)
9. YORK	17	81	(2)	73	(6)	43	(7)	37	(5)	43	(7)
10. HENRICO	18	73	(8)	63	(14)	46	(4)	38	(4)	38	(10)
11. VIRGINIA BEACH	20	81	(2)	70	(8)	48	(3)	40	(3)	33	(14)
12. WILLIAMSBURG	24	71	(9)	69	(10)	42	(8)	21	(11)	37	(13)
13. HARRISONBURG	30	67	(11)	79	(3)	40	(9)	27	(8)	44	(5)
14. ARLINGTON	33	68	(10)	67	(13)	58	(1)	44	(1)	45	(4)
MEDIAN	13	74		71		42.5		29		41.5	

SOURCE: Virginia Department of Education (1993).
NOTE: All variables in table are percentages. Rank for each variable in parentheses. *Too few cases in denominator to be reliable.

school performance data that they could use to grasp the level of performance of their school (Level 1 graphicacy), view norms or patterns in performance data of other schools (Level 2 graphicacy), and compare their school to other schools on individual measures and overall performance (Level 3 graphicacy).

In the next three sections of this chapter, I focus on a specific type of problem that involves looking at multiple cases graphically. The first section deals with multiple cases and one variable. In this situation, we are often interested in grasping the variation and distribution of all the cases to get a sense of the whole. Some of the graphical displays allow for the identification of specific cases by labeling, others do not. The second section presents a two-site or two-case series of graphical comparisons. These two-site examples are divided into two types: the comparison of the distribution of one variable across two sites and the comparison of multiple variables across two sites or cases. The third section presents an example of multiple-case, multivariate comparisons. This is a common situation for applied researchers that has had few successful resolutions. Therefore, some research on the effectiveness of two solutions is presented also.

COMPARING MULTIPLE CASES ON ONE VARIABLE

Many applied research projects collect data on multiple cases. Many national policy studies report data by state and by region. State studies ranging from studies of highway financing to incidence of child abuse frequently report by city and county. These data are often used by state and local policymakers and their staffs to graph local trends and problems. Understanding the location and dispersion across all the cases is often a necessary first step in understanding their own situation.

In Figure 4.1, the 1989-1990 average teachers' salary data for the 133 school districts in Virginia are presented. In Figure 4.1a, the standard histogram or density plot is displayed. The *Y* axis shows the proportion of the cases in each bar. The bars are set to $500 intervals. It is important to override the software default procedures that set the intervals for the bars to make them meaningful widths for both the data and the viewers' framework for analyzing the data. Approximately 5% of the districts appear to have average salaries above $34,000. The histogram has been overlaid with a curve that represents the normal distribution. This aids

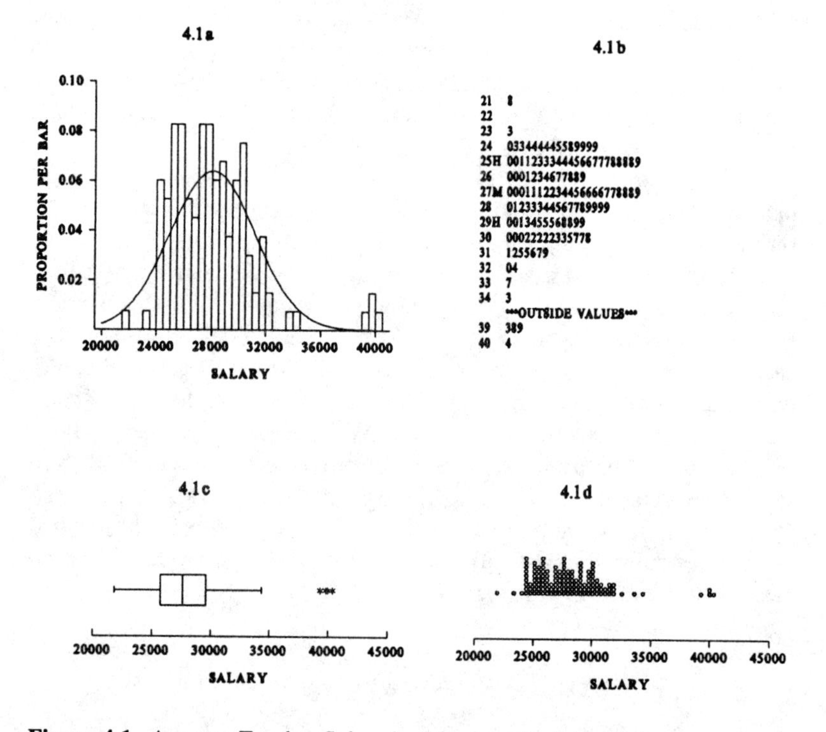

Figure 4.1. Average Teacher Salary by District, Virginia, 1991

our detection of the four outliers that pay the highest salaries, the overly thin tail of lower salaries, and the thick middle, where most of the districts reside.

Figure 4.1d shows a variation on the histogram in which the bars are replaced by a dot that represents each district. Because each dot represents one district, it is easy for a nontechnical audience to grasp the graphical symbols. An axis with tick marks and scales is not needed when this scale is explained. This graph highlights the four extremely high districts and three others that also stand out on the high side. In addition, attention is drawn to the two cases that are separated from the others at the low end of the salary scale. The clump in the middle dominates the graphic.

Figure 4.1c is a significant departure from the previous two displays. Rather than plotting the individual cases, the box plot, based on Tukey's (1977) box and whisker plot, plots mostly summary information. The outside of the box represents the hinges, or the 25th and 75th percentile of the distribution of teachers' salaries. In this case the hinges are $25,755 and $29,605. The center line of the box is the median, $27,643. The lines are drawn to the nearest data point inside a range +/– 2 times the interquartile range. The interquartile range in this case is $29,605-$25,755 or $3,850. Individual data points are plotted using * symbols for values beyond the median +/– 3 times the interquartile range. The box plot quickly shows where the bulk of the data and the median lie, the symmetry of the distribution, and the four cases that stand out on the high end. The downside of using this plot is the amount of explanation involved when presenting the graphic to audiences unfamiliar with the median and interquartile range.

The other graphic, Figure 4.1b, another Tukey (1977) innovation, presents the information available in the other graphs in one graph. This graphic is known as the stem and leaf plot. Because the graphical symbols represent data, Tufte (1983) uses it as an example of a multi-functioning graph (p. 140). On the stem, the vertical axis on the left side of the panel, appear the first two digits contained in the data, in this case 21 through 40. These indicate the data ranges from the lowest average salaries in the $21,000s to the highest average salaries in the $41,000s. The third digit for each district is shown as part of the leaf. On the top line, the first digit listed to the right of the stem, 8, represents the salary of Highland School District, $21,852, the lowest average salary in the state. The first three digits (remember Wainer's recommendation to round the number of digits) of each district's average salary is so displayed.

In addition, the horizontal bar containing the median and the two hinges are indicated by M and H on the stem. Outliers are indicated as outside values. Although packed with information, the need for explanation of this graph to a nontechnical audience is a significant price to pay for the additional information. All in all, Figure 4.1d presents an extremely viable option for most single-variable, multiple-case displays (for a larger version of this graph, refer to Figure 1.3). As additional information, labels for the outliers or a particular district can be added, or norms such as the median or mean, as is done in Figure 1.3.

When it is important to identify the cases, another type of display can be used. In Figures 4.2a, 4.2b, and 4.2c, bar charts and a dot chart are presented—in Figure 4.2a, a bar chart with the districts sorted in descending order from top to bottom. Bertin (1967/1983) recommends this sorting or "repartitioning" to simplify the image in a way that results in "a quantitative series which tends toward a straight line" (p. 104). The graphics in all three figures are oriented as horizontal bars to facilitate reading the labels. The ability to read the labels also

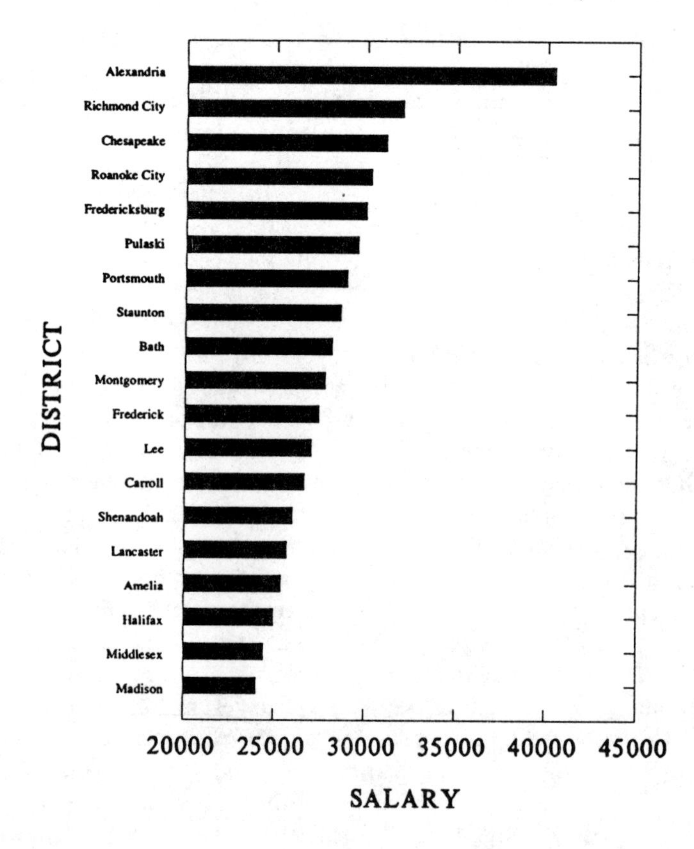

Figure 4.2a. Average Teacher Salaries, Virginia, 1991

depends on the number of units in the display, the size of the display, and the type font used. The page size for this text permits a maximum of approximately 20 units if the type font is sans serif or very plain. For the display area on an 8½-x-11-in. page, about 40 units can be displayed. Type size that is adequate for high-quality laser printing will wash out with copying or if transferred to a transparency for an overhead. A larger type face is required, and fewer units can be displayed. Also, legibility on the computer monitor does not imply legibility on the printed page. When individual units are important, examine the printed graph, checking for legibility, before finalizing it.

As noted earlier, Cleveland (1984) makes a interesting point concerning bar charts such as Figure 4.2a: The length of the bar does not convey

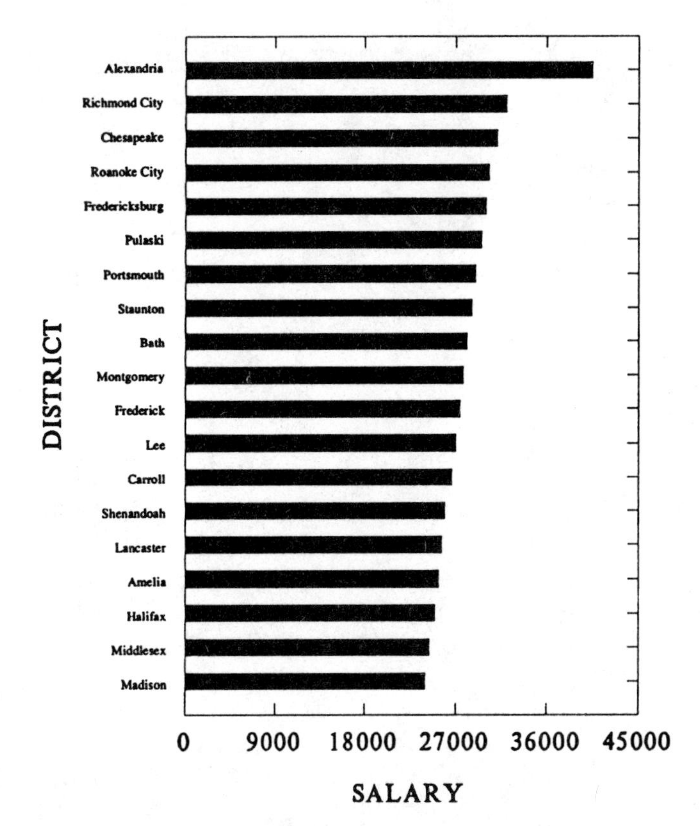

Figure 4.2b. Average Teacher Salaries, Virginia, 1991

the value of the district; it is rather the endpoint of the bar that does so. If the length of the bar is used to judge the percentage differences between cases, the conclusions would be wrong. For example, it appears in Figure 4.2a that in Alexandria the average teacher's salary is twice that in Roanoke City. Actually the average in Roanoke City is approximately $10,000 less than Alexandria's $40,000 average. Figure 4.2b offers one possible remedy for the confusion that could ensue. In Figure 4.2b, the scale of the *X* axis is set to make the minimum point 0. This eliminates the confusion between length and position of the end of the bar in making comparisons. Cleveland (1984) ends up recommending the dot chart solution, which is presented in Figure 4.2c. The zero-base solution, Figure 4.2b, appears to be the preferred solution in this case. However, the scale minimizes the visual perception of differences. For fewer data points, I find Cleveland's (1984) dot chart useful.

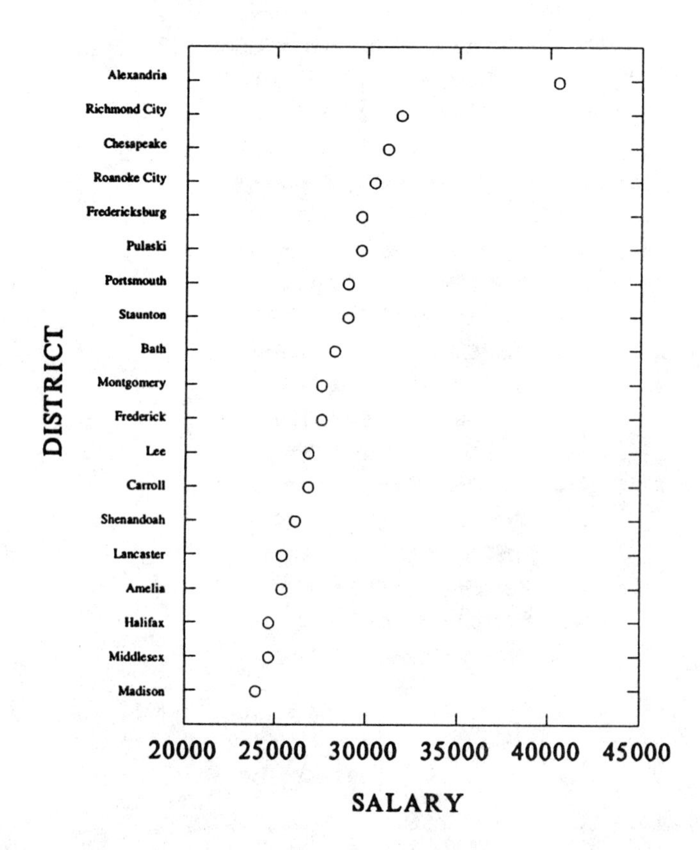

Figure 4.2c. Average Teacher Salaries, Virginia, 1991

These figures illustrate an important point with respect to graphicacy. All three figures can be used equally well for the task of Level 1 graphicacy—reading the salary averages for each district. The descending order of the districts invites the retrieval of the trend across the districts reasonably well in all three, although Figure 4.2a may exaggerate the differences. However, ascertaining relative differences, the task required in Level 3 graphicacy, is more difficult with both Figures 4.2a and 4.2c than with Figure 4.2b. If the analyst intends to encourage comparison of districts or groups of districts, the latter is preferred. It is interesting to note that a recent ad trying to exaggerate the difference between the product being advertised and other brands uses the truncated base bar chart similar to Figure 4.2a.

COMPARING TWO GROUPS

In some cases, researchers wish to compare two groups. An experiment with one treatment group and a control group is a common example. Another example is the comparison of a single case or site to a norm or typical case. Some of the graphical display types similar to those introduced in Chapter 3 can be used for these purposes. Two groups can be compared on multiple variables using the two-group display in Figure 4.3. In this display, the number of items correctly scored by two groups on a test of graphical comparisons is shown. One group was provided information in a table, the other in a graphic. It is obvious that a higher proportion of the group using the table was able to answer all seven items correctly than those using the graphic. After viewing this graph, one might reasonably conclude that the group using the table answered the questions more accurately than the group with the graph, on average by about one question. The analyst may wonder if this might be due to a particular item or items.

In Figure 4.4, this hypothesis is investigated using the multivariate version of the same graphical display type. In this graphic the two groups are retained, those using a table and those using a graphic. But the bar lengths are proportional to the number of correct responses on *each* item. Three of the seven items appear to be particularly difficult using the graphic (Items 2, 3, and 4). Displaying this information for visual inspection provides more information than the previous graph or a statistical summary of the overall results. By pointing out that three items caused less accuracy with the graphical displays, Figure 4.4 can guide the analyst to investigate the factors that distinguish these three items from the others. The results on the further investigation are presented later in the chapter. The point here is to demonstrate the utility of the graph in guiding the analysis.

The original graphical presentation of these data stimulated more analysis and additional insights about the subject and the effectiveness of data presentation—more analysis and insight than the analysis of variance that was done to determine if the two groups were different. Often the same form can be used with public opinion data and other applied research data. Figure 4.5 shows the responses to a student survey concerning potential issues with relocating university classrooms and faculty offices. Clearly the women students had greater concerns with security than the men. This indicated the need to pay special attention to women's concerns during security planning. The

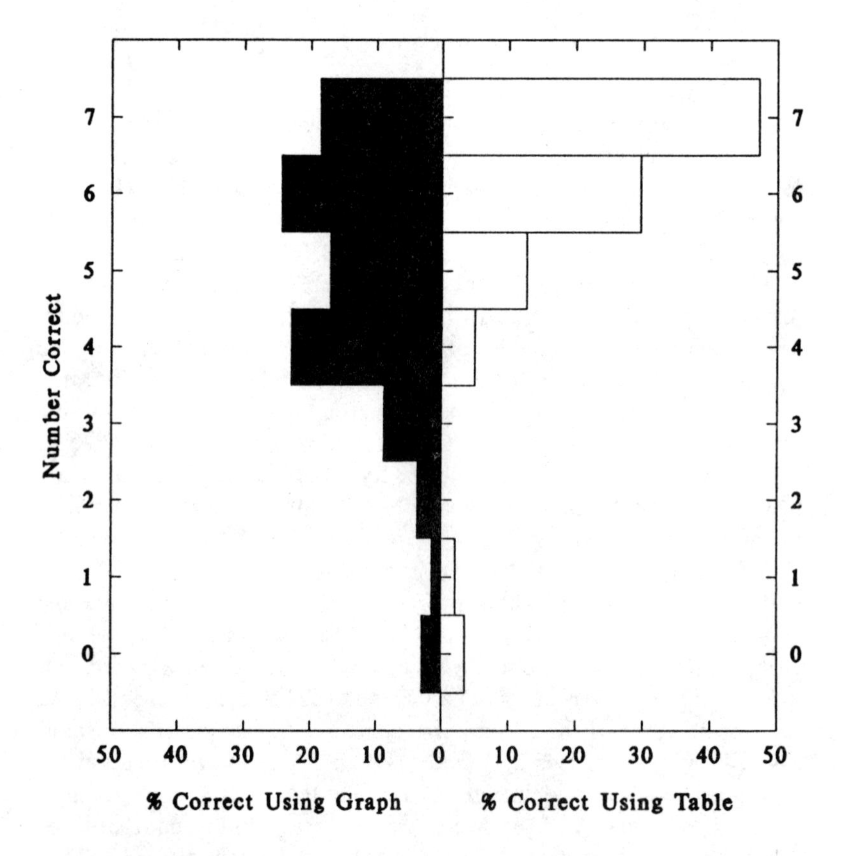

Figure 4.3. Percentage Correct Using Multivariate Graph or Table

graph also shows that parking and building security were major concerns and that students were less concerned about getting to faculty offices than other campus locations—unsurprising results for an urban, commuter campus, but useful for security planning.

Many software programs have routines that allow the analyst to compose these types of graphics with ease. In some packages, routines for displaying age distributions by sex are included. Often these routines will allow the analyst to use other groups (treatment and control, for example) and variables such as test scores rather than age. Some programs can be rigged by changing the values for the group that is to appear on the left to negative values and using a stacked bar graph. In other cases, two density graphs can be aligned along a common spine

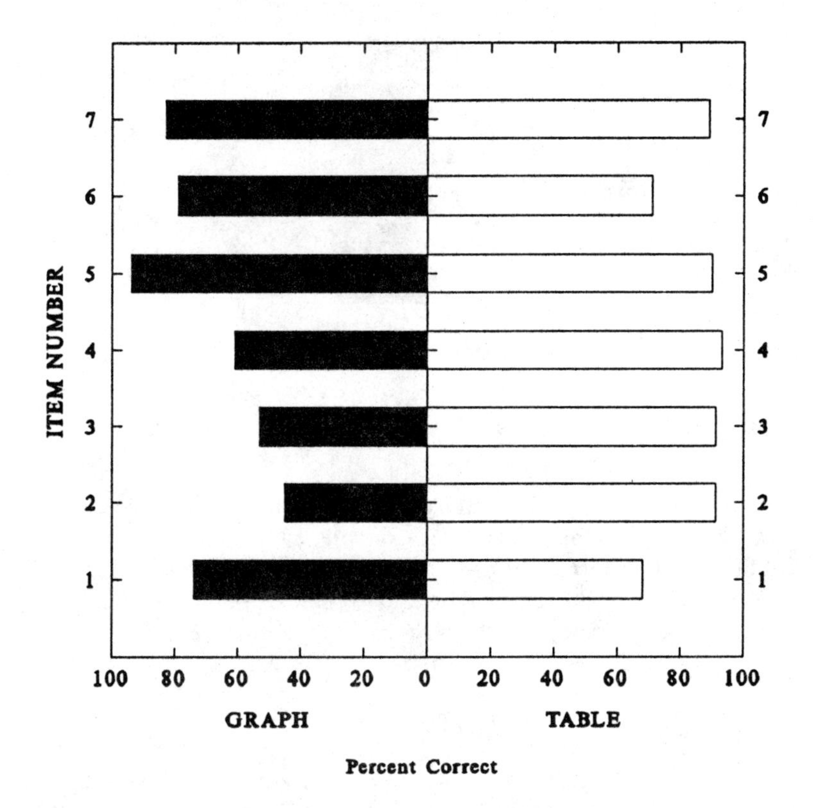

Figure 4.4. Percent Correct for Each Item With Graph and Table

with origin-setting commands and reversing the orientation and order of one graph. Generally, the program can be set up once and saved, then modified for other applications. Developing a standard labeling convention for your customized graphic programs can save a great deal of time in preparing new graphics.

These graphics are obviously useful in allowing viewers to come to their own conclusions and in stimulating more questions and analysis. The graphical technique involved in these last three figures, aligning the responses of two groups along a common vertical spine, is extremely useful in promoting comparisons *both* across items for a group *and* between groups on an item. The technique is limited to two groups and items using the same scale, for example, percentage of correct responses in Figure 4.4. But often more than two groups or cases are involved.

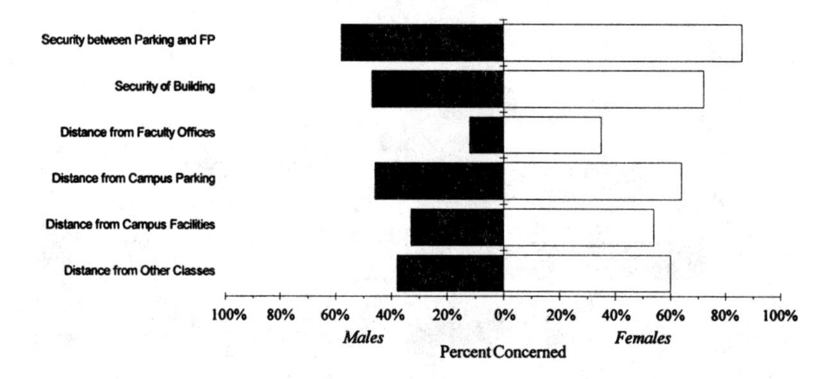

Figure 4.5. Security Concerns in Fairlie-Poplar (Male vs. Female)

Variables of interest to the analyst and the audience are not necessarily measured on a common scale. For example, an audience interested in public school performance indicators may wish to see know test scores, dropout rates, and attendance rates and to compare them graphically. In this case, multivariate, multiple-case graphics are needed.

MULTIVARIATE, MULTIPLE-CASE GRAPHICS

Numerous federal statistical reports present state-level statistical summaries for several variables. Presenting state-level information on morbidity rates by several major causes is an example of numerous reports where data display needs are typically relegated to tables, or in a more limited number of cases, to statistical maps. The U.S. Department of Education's Wall Chart is an excellent example of these displays. Each year through 1990, a Wall Chart has been issued in the form of a 36-×-27-in. table that shows scores and ranking of each state and the District of Columbia on a variety of educational indicators. These data are clearly inaccessible for the purpose of making any but the most limited comparisons, for example, What is the ranking of my state on these variables? Two multivariate displays of these data are included to show a greater use of the data.

The Profile Graph

The profile graph is a multiple-case extension of the display shown in Figure 4.3 and 4.4. Essentially the bars are replaced by points and the variables are shown for each case in a compact format. The Wall Chart data, as described in Ginsburg, Noel, and Plisko (1988), are displayed in this format in Figure 4.6. Using the data obtained from the Wall Chart published in 1989, student performance indicators, SAT scores, and graduation rate were selected for analysis using graphical displays. In addition, a third outcome measure, a SAT score adjusted for the percentage of students taking the test, is included (Powell & Steelman, 1984, 1987; Wainer, 1986).

The profile graphs in Figure 4.6 present the three variables for each state in the Southeast and the District of Columbia. The first scan of the graphs reveals significant differences in overall size of the graphs. Maryland is the largest; the District of Columbia appears to be the smallest. The size is determined by the relative level of performance on the three educational measures: a larger figure corresponds to higher scores. The shape of the figure provides additional information. The three points on the graph represent the unadjusted SAT score, the adjusted SAT score, and the graduation rate from left to right. Maryland, Delaware, and Virginia have a similar shape. In all three, the adjusted SAT score is the highest point. Graduation rate is the second and the unadjusted SAT score appears to be the weakest measure. At a glance, Georgia's scores are low and there is no discernible difference between the scores on any of the three indicators. North Carolina and South Carolina have similar shapes that indicate very low scores on both SAT scores, but relative to those indicators, better graduation rate scores. However, the graduation rate scores fall in the middle of the pack relative to the other states.

The profile graph is an area plot when filled as in Figure 4.6. The viewer can compare overall scores across states, individual scores within a state, and to a lesser extent, individual variables across states. This type of graph can be drawn by connecting the midpoints of bars that represent the scores on individual variables that have been grouped by state and filling in the figure. SYSTAT through its graphical component, SYGRAPH, offers the profile graph as a menu selection in the icon routine.

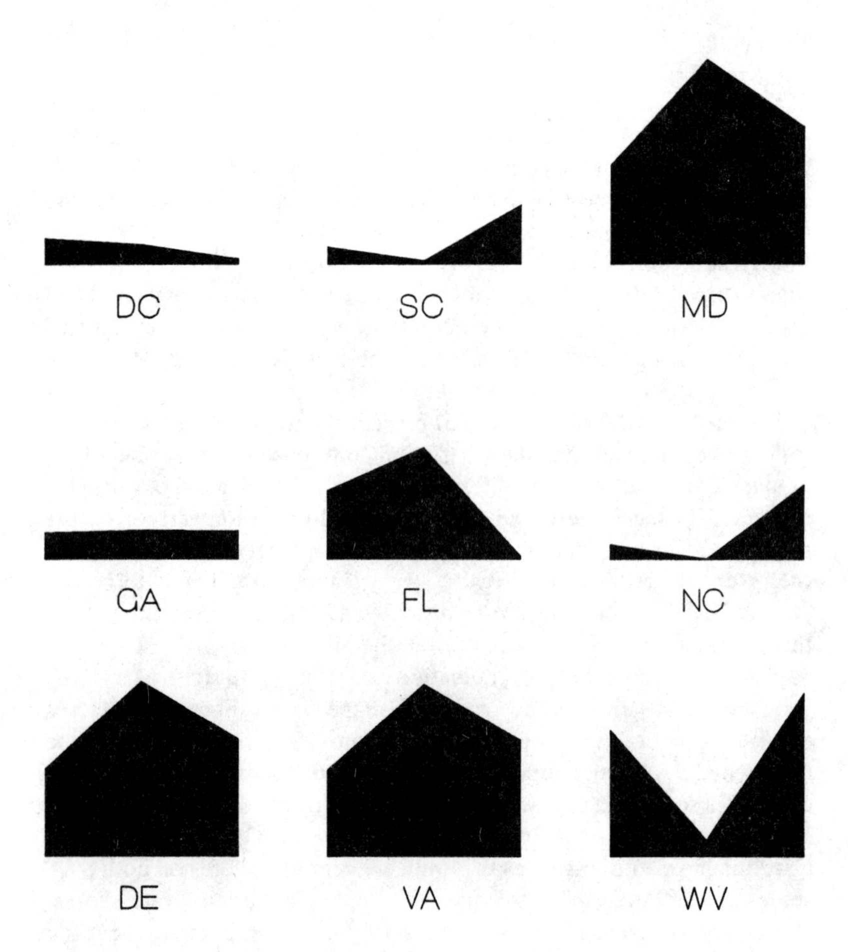

Figure 4.6. State Education System Performance

The STAR Graph

The STAR icon (see Figure 4.7) is one of the most promising graphical displays for evaluation data (Anderson, 1957). STARs are not literally stars, but plots that end up as polygons having as many sides as the number of variables to be plotted; three variables, displayed as triangles, are the minimum. STARs have several advantages. First, they are relatively easy for a nontechnical audience to decode, as has been shown for four-variable plots (Henry, 1993). As these plots are used more often, the form is stored in long-term memory and the decoding process becomes quicker for the consumer of the information. Second, these plots require relatively simple mathematical manipulations of the data and in most cases can be developed to allow the audience to read actual values directly from the display. In Figure 4.7, the minimum and maximum values for each of the axes is presented under the axis label. The actual graphs can be plotted on top of the axes or without the axes, as is the case in Figure 4.8. If the axes are plotted, the scales can be recorded on the axes and the plot will facilitate the first level of graphicacy, discerning levels. This level of graphicacy is sacrificed with the profile graph and with the STAR graph when the graph is not plotted on the axes.

Third, the STAR icons involve multiple perceptual tasks, including perception of length, perception of angles, and perception of area (Cleveland & McGill, 1984; Simkin & Hastie, 1987). The length perception is the distance of the point of the STAR from the origin. The angle perception task is used in judging the relative value of a particular variable. A more acute angle indicates a higher score on a particular variable for that case. The perception of area is used to compare one case to another on the overall values. A larger STAR indicates higher scores. Using these tasks, STARs can be used for comparing the variable values of one unit, or in the same display, for comparing the values across several units. The shape of the graph and its area are interpretable by a quick scan, even in cases where a member of the audience does not wish to invest the time in using the full potential of the display for comparisons. The STAR icons can be ordered on the page by an important variable to promote specific comparisons and analysis.

Figure 4.7 shows the STAR icon using the three outcome variables from the earlier figure. The axes have been drawn in Figure 4.7 for illustrative purposes; the axes are omitted from the display in Figure 4.8, as they are when using the SYGRAPH software. The three axes, *X*, *Y*, and Z, correspond to the unadjusted SAT score, adjusted SAT score, and graduation rate, respectively. The triangle has been drawn to correspond to the highest value for each of the three variables. For the 50 states and the District of Columbia (DC), the highest average unadjusted SAT score was 1084 and the lowest was 836. The standardized score is given in parentheses. The graduation rate ranged from 58.0% to 90.9%. The adjusted SAT score, which is depicted on the vertical axis, ranged from 51.2 above predictions based on the percentage of students taking the exams to 51.9 below predictions. For most audiences the original scale (unstandardized) should be used on the axes. Because the trans-

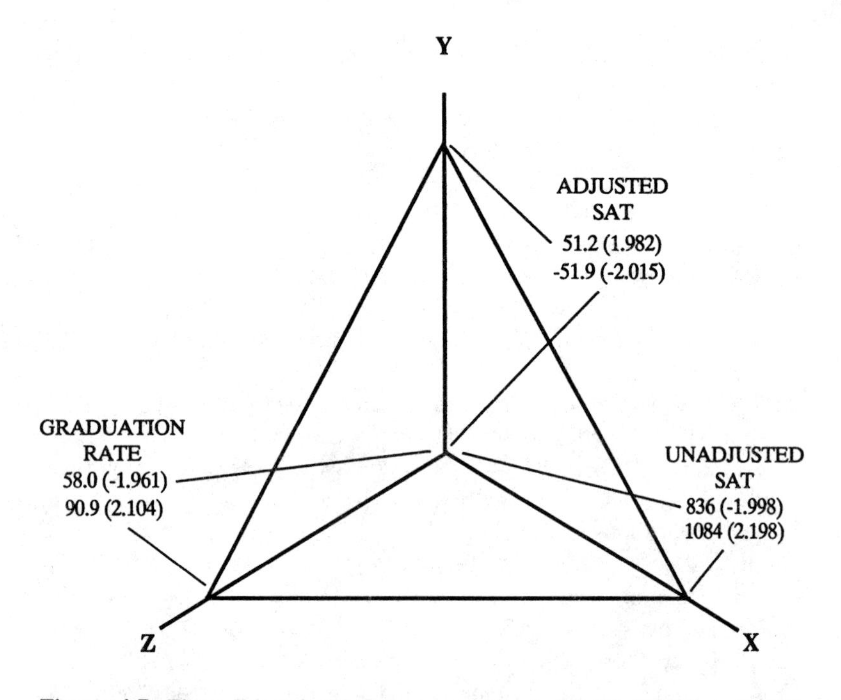

Figure 4.7. Three Educational Outcomes: Adjusted SAT, Unadjusted SAT, Graduation Rate
Minimum and maximum value for each variable on axes (standardized)

formation is linear, the use of the original scale is transparent to the viewer. This makes the interpretation of the values easier and leaves the standardized variables for those who are more interested in the method than the substance.

In Figure 4.8, the performance of the states in the Southeastern region and the District of Columbia on the three outcome variables discussed above are displayed. Area or size of the triangle is an indicator of performance on all three variables. Maryland has the largest overall area, indicating the highest cumulative score, defined as the sum of the three variables. However, the relative similarity in shape and size of Delaware and Virginia indicate that the first-place designation is a distinction without a difference, something that may not be obvious from rankings. In these three states, the adjusted SAT score seems to register the highest performance in intrastate comparisons. Graduation

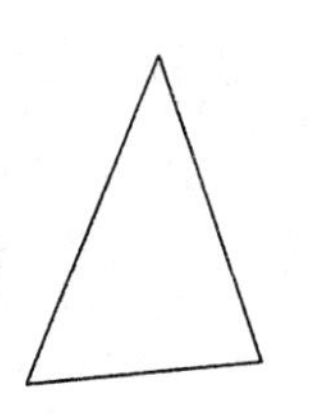

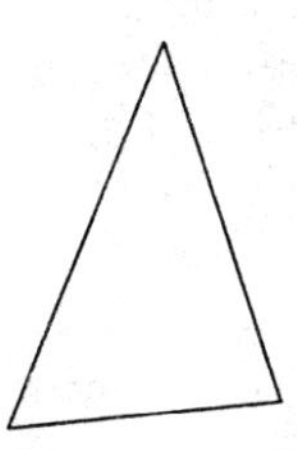

Figure 4.8. Educational Outcomes in the Southeastern States and District of Columbia

falls next, as shown by the lower left corner. All three do less well on the unadjusted SAT, which indicates a high percentage of students took the test. The District of Columbia exhibits the lowest overall performance and its weakest score is the graduation rate on the *Z* axis, which indicates the area where improvement is most needed.

North Carolina and South Carolina have similar performance patterns. Their relative strength is in the graduation rate. However, it does not appear to be a strong score when compared with the three states on the left side of the graph. Florida's weakness is the graduation rate when compared with its other outcome measures and the other states. Georgia displays an equilateral pattern indicating no relative strength or weakness among the three measures. At a glance the size of the Georgia triangle gives an overall indication of weak performance on these three measures. For those interested in a specific state, its internal strengths and weaknesses can be assessed and its performance can be compared with other states using the STAR icon.

Another application using the STAR icon is shown in Figure 4.9, which shows the performance of one school division on four composite indicators of performance. Each of the composite scores is displayed on an axis. The axes can be described by analogy to a standard compass: running from north to east to south to west are the four composite indicators, graduation rate, college preparation, special education, and work preparation. The composite indicators were formed by summing the number of individual indicators on which the school division was rated as within or above expectations. The endpoints of the axes display the number of individual indicators making up each composite. For example, there are eight variables that make up the graduation rate composite and six for work preparation. The number on which the school district performed at or above expectations is defined by the points of the STAR, which in this case is a four-sided figure. The area of the four-sided STAR represents the overall performance of the school district. The district plotted in the example shows strength in the graduation rate, college preparation, and special education. Work preparation appears to need attention. Side-by-side comparisons with similar districts would provide additional useful information for parents, educators, and taxpayers.

Can Audiences for Applied Research Use the STAR Graph?

As a part of a larger experiment concerning the use of graphical displays with evaluation audiences, a STAR graph similar to the one in Figure 4.9 was used as a stimulus. (For details of the experiment, see Henry, 1993). The participants were randomly selected from comprehensive lists of two groups of educators (teachers and school board members) and from the entire population of education print journalists

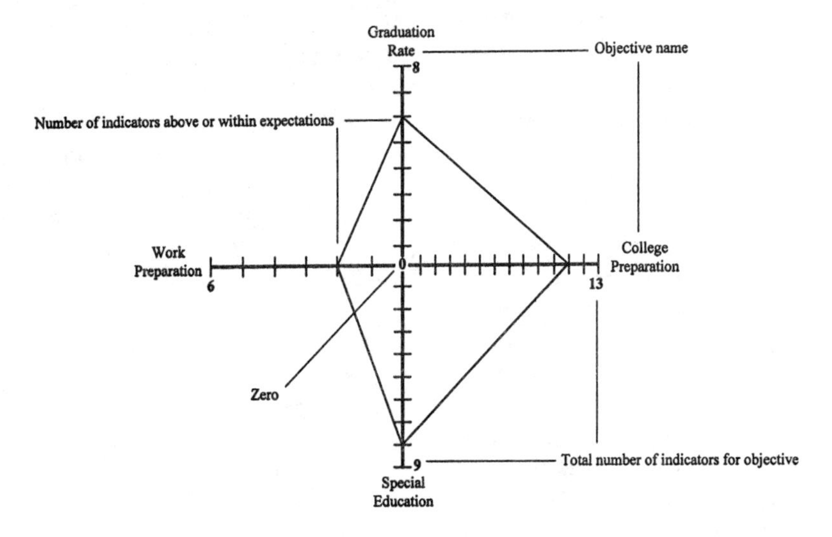

Figure 4.9. The STAR Graph Display for Four Composite Educational Indicators

and school superintendents in Virginia. Each individual was randomly assigned either a table or a graph with the same data. Unfortunately, the data graphic ran over two pages, including two STAR graphs, one of which was printed on the second page. This forced the designer to break with the principle of proximity and turned out to cause the viewers some difficulties. An assessment was made on two criteria: accuracy and audience preference.

Accuracy. On average, the evaluation audiences were able to get one more question right with the table ($x = 5.9$) than with the graph ($x = 4.9$). An accuracy rate of 84% for the tables compares to a rate of 70% for the graph. Figure 4.3 shows a density plot of the accuracy scores for the two multivariate displays. The differences between the two types of displays are statistically significant ($a = .05$).

The results for the accuracy of the multivariate displays were presented in Figure 4.4 and introduced earlier. The individual item comparisons show a pattern that may prove useful for guiding graphic practice. First, as expected, there was no significant difference, statistically or substantively, in the question (Item 1) interpreting the expenditure table common to both display formats (74% and 68% for the graph and table formats, respectively). The greatest differences occurred (91% vs. 45% and 91% vs. 53%) in Items 2 and 3. These two items required comparing displays on two pages, whereas the others could be accurately answered from a single page. This was anticipated to be a significant problem with the design, but the substantive desire to separate two types of indicator categories and the perceptual difficulties with keeping track of more than four axes on a single STAR graph led to the two-page design. This graphical investigation process led to the reinforcement of the proximity principle of graphical displays: *locate important comparisons as close as possible to each other for greater accuracy of comparisons.*

Another important graphic design practice can be wrung from the results on Item 4. This item had the greatest difference in accuracy among the group of items that were able to be answered on one page. Of the participants who had the answer listed in a table, 93% were correct, whereas only 61% of those presented with the graphic were able to answer the item correctly. The graphic display users had to calculate the number of indicators not above or within expectations, a distance that was not included in the graph. To provide a concrete example, a similar question can be asked based on Figure 4.9: In how many work preparation indicators was the district not above or within expectations? In this example, the answer is four, a number not explicitly depicted in the graph. Only 61% completed the task accurately. These data support the warning offered by Stock and Behrens (1991) about "the use of empty space to convey quantity" (p. 18). As Bertin (1967/1983) has pointed out, empty space can accurately convey the null set or zero, but it should not be used to represent a quantity.

Preference. The between-subjects design did not permit a head-to-head comparison of preference between the table and the multivariate graph. Instead, questions were asked about understanding performance of the district, usefulness of the display, and whether the participant liked the display. Results indicated a clear preference for the table. In each case, the difference is statistically significant ($a = .05$). The immediate reaction of the participants, comparing a relatively simple table to a new graphic format, was in favor of the table. It is difficult to say whether the graphs would be found more useful after the original time spent in the decoding process had been spread across several uses of the graph.

The STAR icon shows promise as a useful and versatile technique for displaying evaluation results. Over 80% accuracy in answering Level 3 graphicacy questions seems encouragingly high for those items for which the design was most appropriate. Its usefulness diminishes when more than six outcomes variables are to be displayed, due to the difficulty in distinguishing the individual axes, or when the viewers are asked to view graphs spread over two pages. To use the STAR icon all data must be standardized or in the same original metric. This makes the maximum score on each of the axes of approximately equal length. Also, it is useful to have the values for all cases converted to a scale in which higher values represent "better conditions," which may require a simple transformation.

SUMMARY

Applied researchers often need to display data on multiple cases to allow the audience to grasp the range of differences, and in some instances, the level of specific individual cases. A variety of graphical types are available for these tasks, but each has limitations. Histograms can show the distribution of the data in detail and can be called upon to show the data for two groups. Box plots give a variety of important statistics in graphical form. Text Graph 4.1 shows one situation, when the data are bimodal, that is obscured by the box plot. Profiles and STARs offer reasonable methods for displaying multivariate data. However, audiences are not experienced with these types of graphs. To use them, a researcher must be prepared to pay the price of "educating" the audience in the use of the graph. We all might be better off if these graphs were as familiar to the public as bar charts, but it is hard to be the first to use them. Situations where a verbal briefing will accompany the delivery of the data may be the best way to increase audience attention to more complex graphs such as these.

When considering a graph to display multiple units and multiple variables, a number of steps must be worked through. First, the number of cases or units is important. For two units or groups, there are more graphical options. With more than two cases, the next question is the importance of identifying individual cases. Histograms, box plots, stem and leaf plots, and histograms with dots (all shown in Figure 4.1) are useful univariate summaries for multiple cases. When multiple cases must be displayed with individual cases identified, bar charts with a zero base are useful (Figure 4.2b), but they are limited to one variable. Moving to multivariate displays in which individual cases can be identified increases the complexity of the display. The profiles and

Text Graph 4.1
Bimodal Distribution: Comparison of Histograms and Box Plots

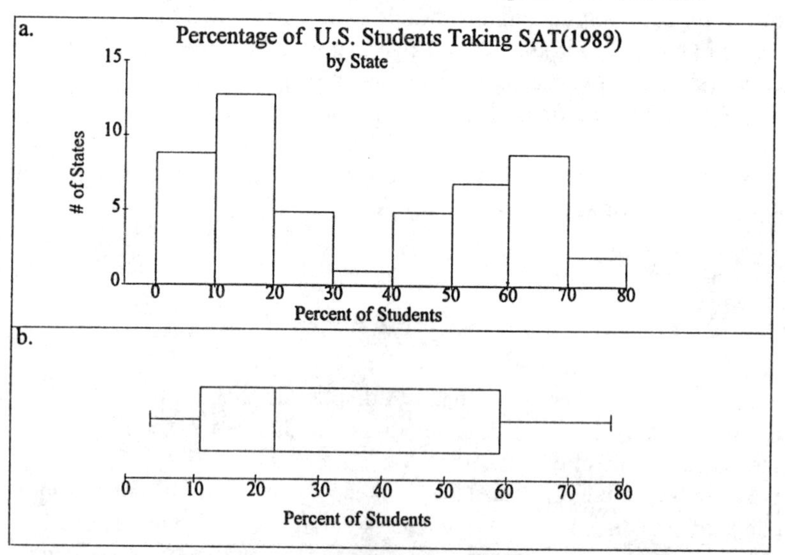

The data presented here are the percentage of students taking the SAT as juniors or seniors for the 50 states. The box plot shows the median and spread with no outliers. The histogram reveals the bimodal distribution that results from ACT-oriented states. The bimodal distribution would have been missed without the histogram.

STARs shown in this chapter are the most abstract displays in the book. They allow the audience to make several types of comparisons that require Level 2 and Level 3 graphicacy. But the trade-off is in Level 1 graphicacy in most cases. It is difficult to read values from these display types, but often it is more important to retrieve the information in a relative sense than a precise number and unnecessary to read the numbers from the chart.

The trade-offs inherent in the choice of alternative graphical displays require the analyst to add some steps to those previously mentioned. Consideration must be given to the level of graphicacy required of the audience to retrieve the information. For Level 1 graphicacy, where the graph substitutes for a table as well as facilitates comparisons, multivariate graphs may be more difficult, often requiring an individual display for each case. Refer to Figure 2.3 for an example. Consideration must also be given to the viewers and their level of exposure to graphical displays. Some individuals seem to immediately attempt to read the levels for the case or group in the display. Others construct comparisons in a quick, first scan. If the viewer is detail oriented, a different graph may be appropriate than for the big-picture viewer. Clearly, the type of data, the level of graphicacy that is sought to be stimulated, the type of display, and the audience must be carefully considered during the development of the graph. Choices must be made about the type of information to be displayed and the type of display. Too often old habits and the ease with which the graph can be prepared dominate these considerations and the usefulness of the display is compromised in the process.

EXERCISES

For the following exercises, use the data in Table 4.1. Begin by entering these data in the software package that you intend to use for graphing the data.

1. Display two school districts on three variables. Three bar charts will be a reasonable way to start.

2. Display all 14 districts' values on the percentage of students in the eighth grade above the 75th percentile on a standardized test. Include the district name for all districts. What do you learn from the display when it is ordered by the percentage of students eligible for free and reduced-price lunch? Use the same variable for a box plot, a stem and leaf plot, and a histogram.

3. If you have software available for a STAR graph, select four variables and design a graph. If your software does not have this routine, graph four variables on a bar chart (remember to standardize the data) for each school district, remove the spaces from between the bars, and fill the bars. This should resemble the profile graph.

5

Trends

The most common type of graphical display is the trend line or time series graph (Tufte, 1983). Trend graphs usually have a measure of time, either years, months, weeks, or days, on the horizontal or *X* axis and one or more dependent variables on the vertical or *Y* axis. In Figure 5.1, the crime rate for the United States is plotted since 1960. The crime rate in the plot is measured as the number of crimes reported in the FBI's Uniform Crime Reporting system per 1,000 people. We can use the graphic to answer questions at each of the three levels of graphicacy:

Level 1: What was the crime rate in 1970?

Level 2: What was the trend in the crime rate between 1970 and 1980?

Level 3: How is the crime rate between 1970 and 1980 similar to or different from the crime rate between 1980 and 1990?

Each of these questions can be answered from the figure, but a more efficient design could be developed if a specific question was chosen. Trend graphs have become so commonplace that they are often used without consideration of the purpose and message for the graph. Too often they become clutter rather than information for the audience. I begin this chapter with some background about trend graphs and then consider some of the major issues in the design of time series graphics.

PLAYFAIR STARTED THE TREND

Trend graphs were the first to use the *X-Y* plane to represent data that do not have a complete and direct physical analogy. Earlier graphical depictions were maps and charts in which the plane of the paper itself is intended as an accurate physical representation of spatial relationships. Maps simply alter the scale of physical features in representing them. Trend lines change the *X-Y* plane from the two-dimensional representation of space to a more abstract, symbolic representation of the level of a variable at set intervals of time. Time is charted across the

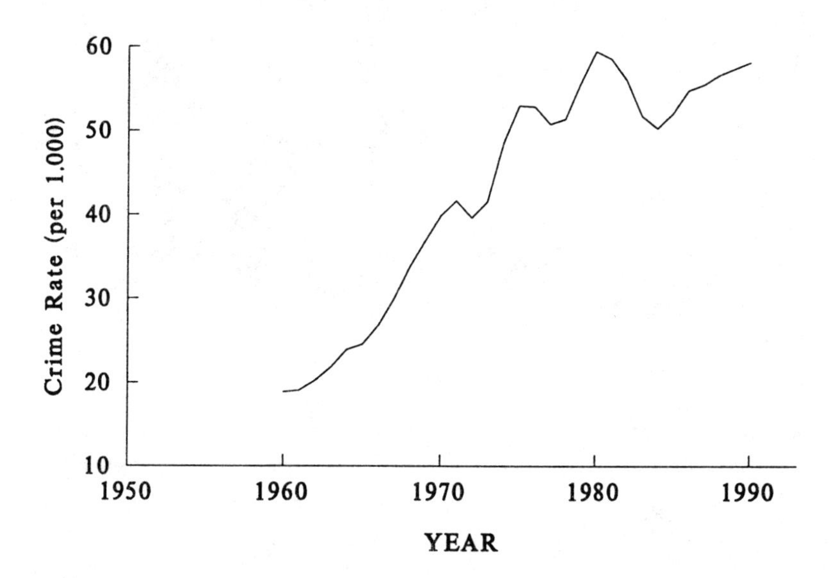

Figure 5.1. Reported Crime Rate in the United States
SOURCE: FBI (1991).

horizontal axis, moving forward from left to right. This convention seems to have stood for nearly 1,000 years, since the time of the first known graphs of the movement of the planets and the sun (Funkhouser, 1936). The trend line has been useful in nearly every scientific field and in communicating data to a variety of audiences. In addition, the development of plots of two variables on a plane set the stage for the Cartesian graph, an entirely relational graphical form that frees the graph designer from the constraints of time and space to create wholly symbolic representations.

Time series graphs began to be popularized in the late 1700s, according to Beniger and Robyn (1978). Most graphical historians attribute the popularization of the innovation of time series with economic and social data to William Playfair in his *Commercial and Political Atlas* (1786), where the first bar chart was also presented. The time series that Playfair produced are still cited as elegant examples for current practice (Cleveland & McGill, 1984; Fienburg, 1979; Tufte, 1983; Wainer, 1992). His numerous examples of imports and exports highlight the trade imbalances from year to year with clearly labeled scales on both axes, nice proportions of length-to-width dimensions, and a minimum

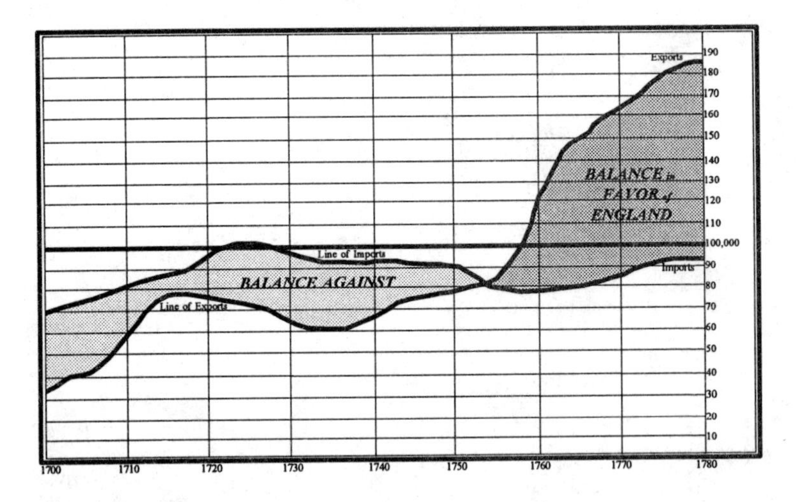

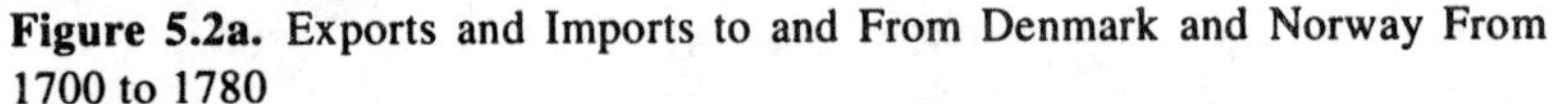

Figure 5.2a. Exports and Imports to and From Denmark and Norway From 1700 to 1780
SOURCE: Derived from a reproduction of an original work by William Playfair found in Tufte (1983).
NOTE: The bottom line is divided into years, the right-hand line into £10,000 each.

of distractions from the data, as can be seen in Figure 5.2a. However, it is difficult to accurately compute the year-to-year imbalances by eye.

To focus more directly on the balance of trade, I have replotted one of Playfair's graphs using computer software, but staying as true as possible to the original in Figure 5.2a. Figure 5.2b plots the differences directly, thereby focusing the viewer on the trade balances rather than the levels of imports and exports. The original presents more data than the latter, in which the levels of imports and exports are lost, but the latter is more efficient in presenting the balances. Playfair's later work was multidimensional, plotting three related time series on one graph. It is little wonder that trend graphs have become so popular given the elegant presentation of the researcher who first popularized them.

Perhaps the basic simplicity of the trend graph has caused it to be so abused in the hands of graphic designers over the years. With important data, some elementary mathematics, and graph paper, time series graphs are readily constructed. However, an overemphasis on entertainment and an underemphasis on displaying data to facilitate the exchange of information seems to have spawned numerous faulty designs. After slogging through numerous examples of time series charts trying to

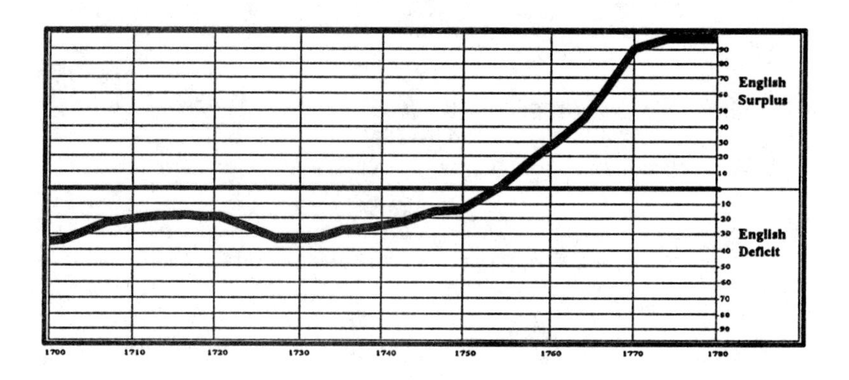

Figure 5.2b. Balance of Trade Between England and Denmark and Norway from 1700 to 1780
NOTE: The bottom line is divided into years, the right-hand line into £10,000 each.

retrieve the data and trends supposedly presented in them, I believe the straightforward advice of Edward Tufte (1983) is important in attaining graphical competence: "Above all else show the data" (p. 92).

With this credo in mind, I discuss the most important elements of presenting data in time series graphs, beginning with how much data to graph in a single series and including the use of smoothed lines to indicate trends. Sizing the graph is the topic for the next section, followed by details of the graph such as labels, grid lines, and scale indications. Finally, I close the chapter with the topic of multivariate time series.

GRAPHING TIME SERIES DATA

For the graphs in this chapter I will use three data sets that were generated by applied researchers and originally presented by them in graphical form. The first series, presented in Figure 5.1, shows the crime rate in the United States since 1960. The second series, in which annual Medicaid expenditures in Virginia are compared, is depicted as originally published in Figure 5.3. The figure has been redrawn to accurately reproduce the original graph. Quite obviously this multivariate time series graph is taking on numerous tasks—perhaps too many tasks for one graph. You can judge the success of the graph by using it to answer questions at each of the levels of graphicacy, for example:

Level 1: What was the expenditure for pharmaceutical services (drugs) in 1989? What about inpatient hospitals in 1985?

Level 2: What has the trend in mental health service expenditures been since 1988?

Level 3: Which expenditure category has grown the most over the past 5 years? Which expenditure category is responsible for the drop in overall expenditures in 1983?

After careful study the viewer may ferret out that inpatient hospital expenditures dropped significantly in 1983, largely causing the overall drop. However, the viewer quickly realizes that none of these questions can be answered confidently from the graph. Some of the data are hidden and some of the categories are hard to follow from year to year. Although the graph has a contemporary, professional look, it is not designed to reveal the data it contains. Three specific design choices detract from this data display. First, the addition of the 3-D perspective distorts judgments of bar height. The grid lines run downhill at acute angles to vertical and make it difficult to judge the height of the bar, if you could even determine whether the front or back of the bar conveyed the data. The use of bars rather than lines has necessitated the use of a pseudo third dimension. No other arrangement could avoid hiding some of the bars and the pseudo 3-D has not been entirely successful. Finally, the decision to separately plot the five categories of expenditures and the total on one graph created scale differences that make graphing difficult. Alternative graphical designs with solutions to these problems are presented later. The alternatives allow the questions presented above to be answered confidently, although not all alternatives work equally well for all the questions (see Figure 5.13 for four alternatives).

The third data set is the inmate population in Virginia jails since 1983. This series is recorded monthly, yielding 109 data points to be plotted. The top panel in Figure 5.4 shows the plot of the full data set. A casual look at the graphic shows some fluctuations and a definite trend. Forecasters traditionally break down a data series of this type into three or four components, depending on whether a cyclical component is needed (Wheelwright & Makridakis, 1980):

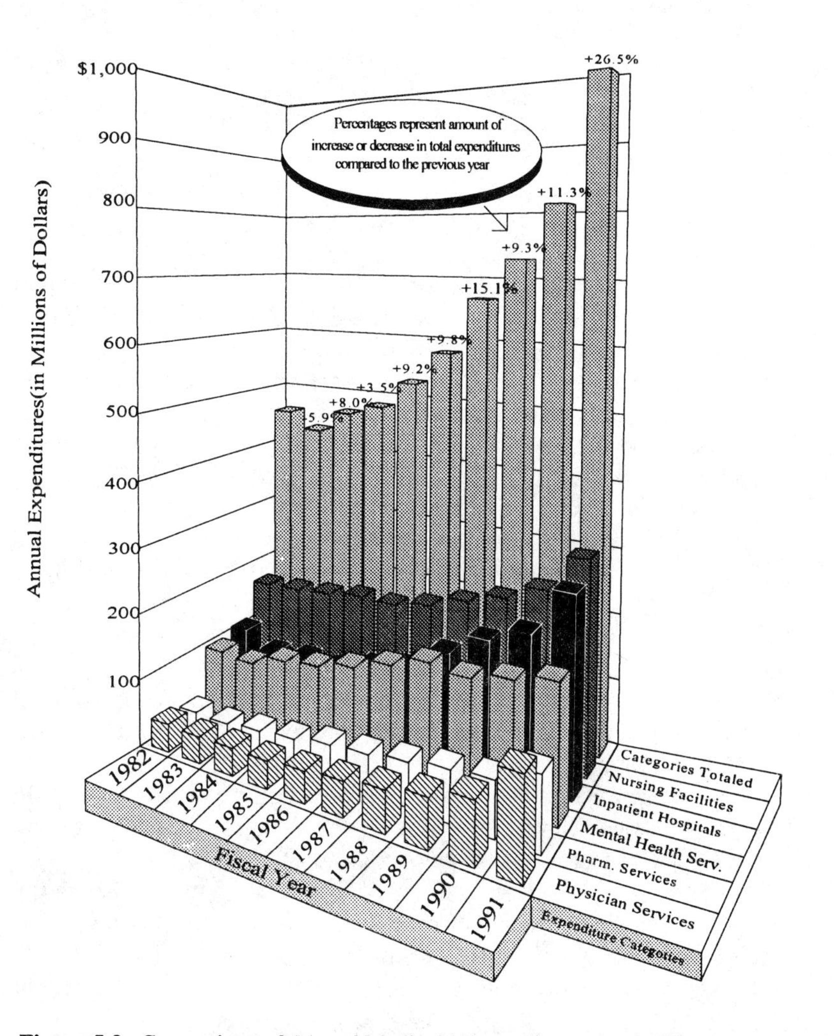

Figure 5.3. Comparison of Annual Medicaid Expenditures for the Five Largest Medical Care Categories, FY 1982-FY 1991

SOURCE: JLARC staff depiction of Department of Medical Assistance Services medical care expenditure worksheet, FY 1982-FY 1991, derived from unaudited financial statements. This chart is a reproduction of an original graphic found in the Joint Legislative Audit and Review Commission (1992, p. 36).

Seasonal component: Variation that occurs in regular intervals in the data series (usually every 12 months for monthly data or every 7 days for daily data)

Trend component: The long-run, linear movement in the data

Random component: The chance fluctuation in the series

Cyclical component: Variation that runs over a longer number of periods than the seasonal component, such as economic cycles

The seasonally adjusted data are plotted in the middle panel in Figure 5.4. The seasonal adjustment is computed by calculating how much the average of each month is above or below the average for the entire series. For example, December's average is usually low because many inmates in local jails are granted holiday furloughs to spend time with their families and maintain their relationships with members of the community. This also aids in reducing jail populations during times when jail employees wish to spend time with their families. Therefore, a seasonal decrease in the jail population is expected. The seasonal adjustment for December is 97.8%, indicating that the number of inmates in jails during December is about 2% lower than the average for the other 11 months. The seasonally adjusted series divides the actual series by the seasonal adjustment for the appropriate month.

The seasonal adjustment for jail populations in December is a modest net reduction, as might be expected, indicating that the number of holiday furloughs is low or that the furloughs are offset by increases in admissions to local jails or longer backups in the courts that result in longer pretrial jail stays. The seasonal adjustment of this series removes only a modest amount of fluctuation, as can be seen from the slight change between the top and middle panels in the figure. Although the effects of removing seasonal trends were modest in this case, the first step in a series simplification should be a deseasonalized plot. The deseasonalized trend can be computed by most software packages that have time series routines for analysis or forecasting. In this case, additional time series plots, such as the number of jail admissions and the size of the pretrial population, would help to answer the questions about the modest seasonal adjustment for the December jail population.

The series is again plotted in the bottom panel in Figure 5.4, this time with the trend removed. The data have been detrended by removing the linear trend from the series, that is, subtracting out the predicted value of a regression line fit to the series. These data show the increase in populations was slower over the first 5 years of the series, followed by large fluctuations above the level of the overall trend. Each of the plots

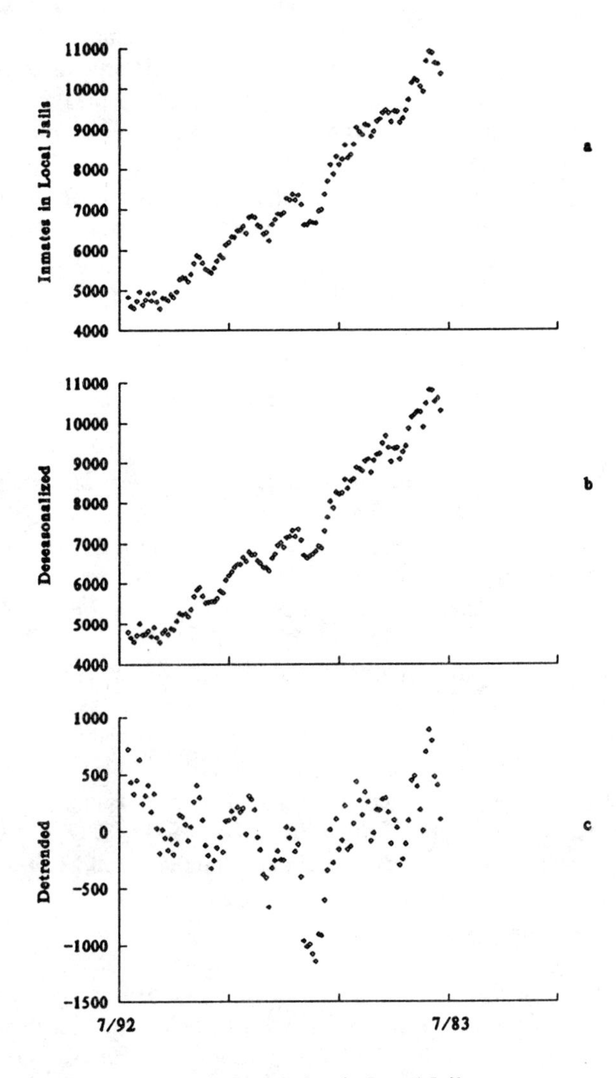

Figure 5.4. Virginia's Inmate Population in Local Jails

gives a perspective on the data that is analytically beneficial. In later sections we will examine ways to improve time series plots for various questions related to the level of graphicacy.

These three data series have characteristics that are common to many time series data used for applied research. The crime data are a long data series containing many short trends and a fair amount of fluctuation over time. The Medicaid data are a relatively short annual series, but important in their implications for the availability of health care, the impact of this entitlement program on governmental expenditures, and understanding its various components. The third series, jail populations, contains numerous data points and requires some simplification, as we have already seen, to focus our attention on some salient changes and relationships. From these series we can extrapolate good practices to other applied research data series.

HOW MUCH DATA?

The first point in deciding how much time series data should be graphed is to make sure that the intervals being graphed are consistent throughout the series. In a recent publication of the U.S. GAO (1992), the deficit was graphed as a percentage of the gross national product (GNP) by decade. The graph has been redrawn with the pseudo 3-D perspective, exactly as it appeared in the original (Figure 5.5). The final interval contains 8 rather than 10 years of data and mixes projections and actual data. As the data appear to be averages, at first blush, this decision appears less problematic. The graphic portrays a deficit that is projected to go down in the 1990s. But the current year, 1992, is expected to be the highest ever, 6.3%, and the deficit is expected to rise from 1995 through the end of the decade (U.S. GAO, 1992, p. 41). Use of unequal intervals misleads the audience, and the pseudo 3-D perspective further confuses the representation. Intervals must be equal for time series graphs to accurately represent the data.

In assessing this graph, we also must consider the method used for the "averaging" calculation. Because the data are percentages, the sum of the deficit for time period, in this case 10 years, should be divided by the sum of the GNP for the corresponding years. This calculation gives the most weight to the largest years, usually the end of the decade, due to inflation. It is not clear that this procedure is followed here or whether the percentage deficits are averaged (the averages summed for 10 years and divided by 10). In any event, the projected years, fiscal year 1992 through fiscal year 1997, would have the greatest weight in the calculation, if it is done correctly. But the years with the highest projected deficit, 1998 and 1999, are left out of the graph. Further,

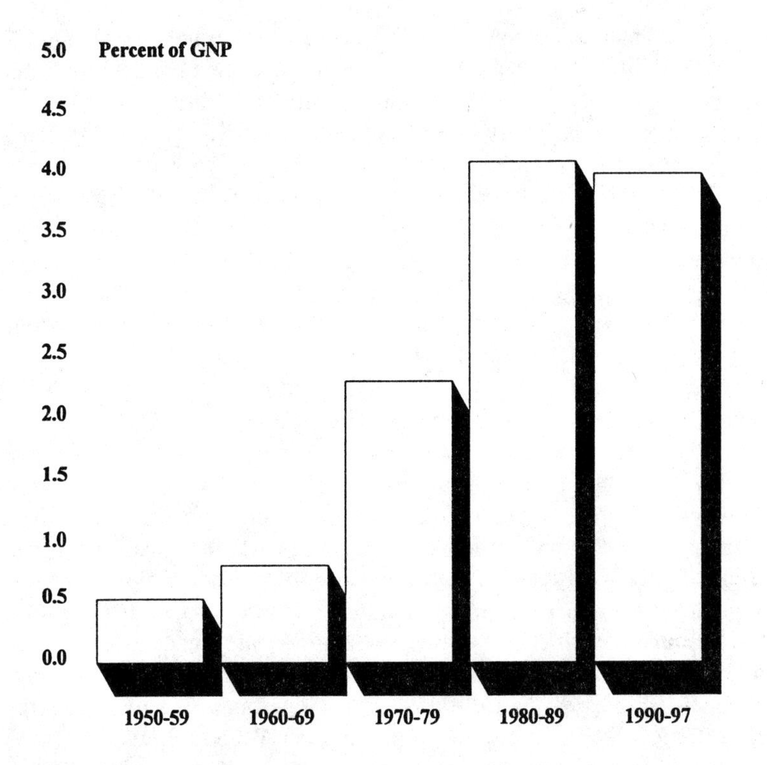

Figure 5.5. U.S. Budget Deficit Projections, 1950-1997
SOURCE: *Budget of the U.S. Government,* 1950-91; CBO projections, 1992-97. This chart is a reproduction of an original graphic found in U.S. GAO (1992, p. 26).

because the deficit is expected to be rising from 1995 through the end of the decade, the years that are projected to be the highest and exert the most influence on the decade averages are left out. The reason for leaving out these years is inexplicable, as they are plotted in a later graphic. This graph gives the impression that the deficit will begin to decrease during the decade, when in fact it is growing and the text emphasizes the deficit's "resistance to reduction measures" (U.S. GAO, 1992, p. 26). The graph counters the author's point, rather than complementing it. The deficit data represent the case where too little data are graphed, summarizing the information by decade is unwarranted, and unequal intervals distort the data and diminish the credibility of the information.

The next example is the all too typical case of a time series using all of the available data and a default plot. A plot of this type depicting the jail population series is shown in the top panel of Figure 5.6. These data seem to have too much fluctuation to imprint an image in short-term memory that goes beyond an overall upward trend. This graph does not allow the viewer to ascertain the differences in the rates of increases from one group of months to another. The trend can be reinforced and specific time periods analyzed by drawing a linear smoothing line, as shown in the middle panel of Figure 5.6. For this plot the line is restricted to the range of the data to avoid the appearance of forecasting or backcasting the data using this simplistic method. In general, it is good practice to avoid extrapolating beyond the range of the data with a simple linear smoothing unless you have a good reason for doing so. However, the default option for most software usually draws the line beyond the range of the data to the edge of the graph.

From the graphic we can see that the jails across the state have had to accommodate about 690 additional inmates each year during this time period. In the early 1980s, the growth appears insignificant; in the middle of the series a drop in jail populations is evident; and in the most recent months the increase seems greater than the linear trend for the entire series. We are not sure that the trend is linear. Cleveland (Chambers, Cleveland, Kleiner, & Tukey, 1983) has developed another type of smoothing using locally weighted sum of squares (LOWESS). LOWESS computes a curve that is dependent on the predicted values from a weighted average of the Y values that lie near each X value. A LOWESS curve follows the trend in data from one period to the next and gives the graph viewer a good idea if the data are linear. The LOWESS curve for this series is shown in the bottom panel of Figure 5.6. The curve appears to be close to linear, but the slope is increasing somewhat over time, given the convex appearance. This curve fits both ends of the series better than the straight line regression. LOWESS is an optional smoothing method in the SYSTAT/SYGRAPH software and is also useful in judging the linearity of bivariate relationships, as is shown in Chapter 6.

In addition to drawing lines to orient the viewer to the overall trend, the data themselves can be smoothed. The seasonal adjustments shown earlier are an example of smoothing to eliminate some of the variation in the data. In Figure 5.7, the same series is displayed three ways. The original data appear in the top panel. Then the data have been converted to a four-period moving average. The moving average is calculated by taking the average of the first four observations in the time series for the first value, then removing the first observation and adding the fifth

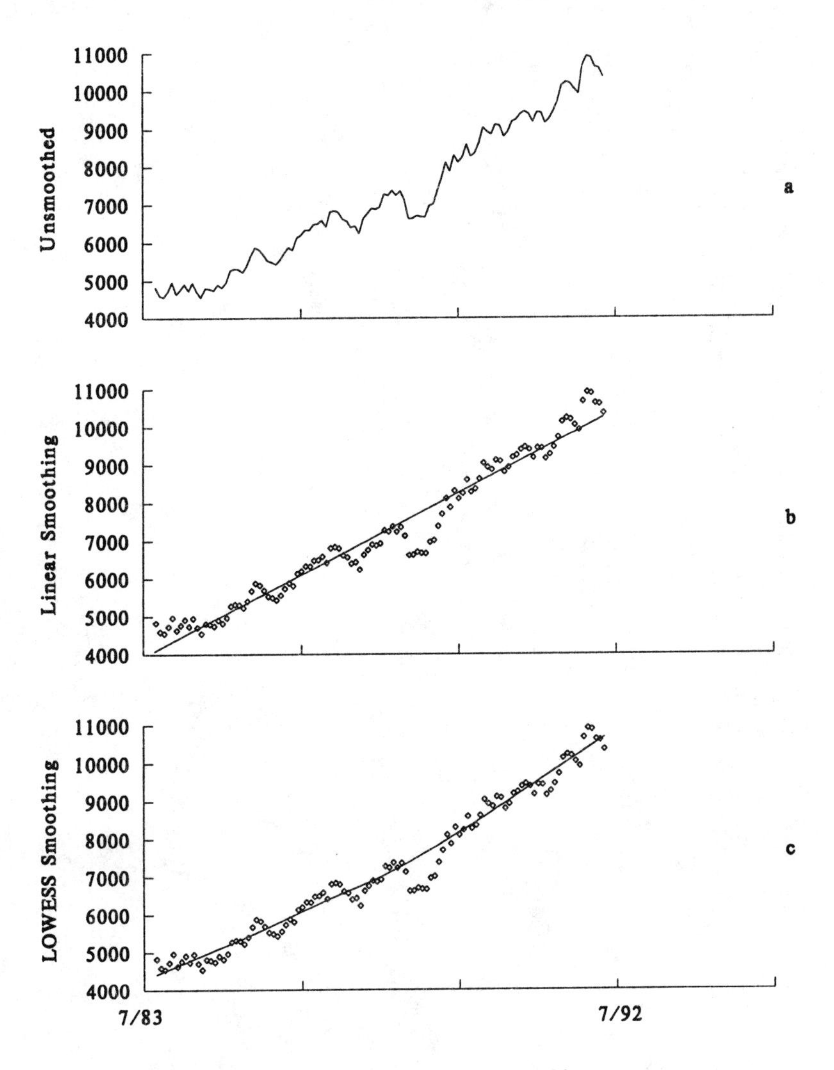

Figure 5.6. Inmate Population in Local Jails: Smoothing

in the ordered series and recomputing the average. This process continues, deleting the earliest observation and adding the next data point throughout the entire series. The moving average plot focuses our attention on the two distinct series: the ripples in the first 60 to 70 observations, and the increased slope of the trend thereafter (see middle panel in Figure 5.7).

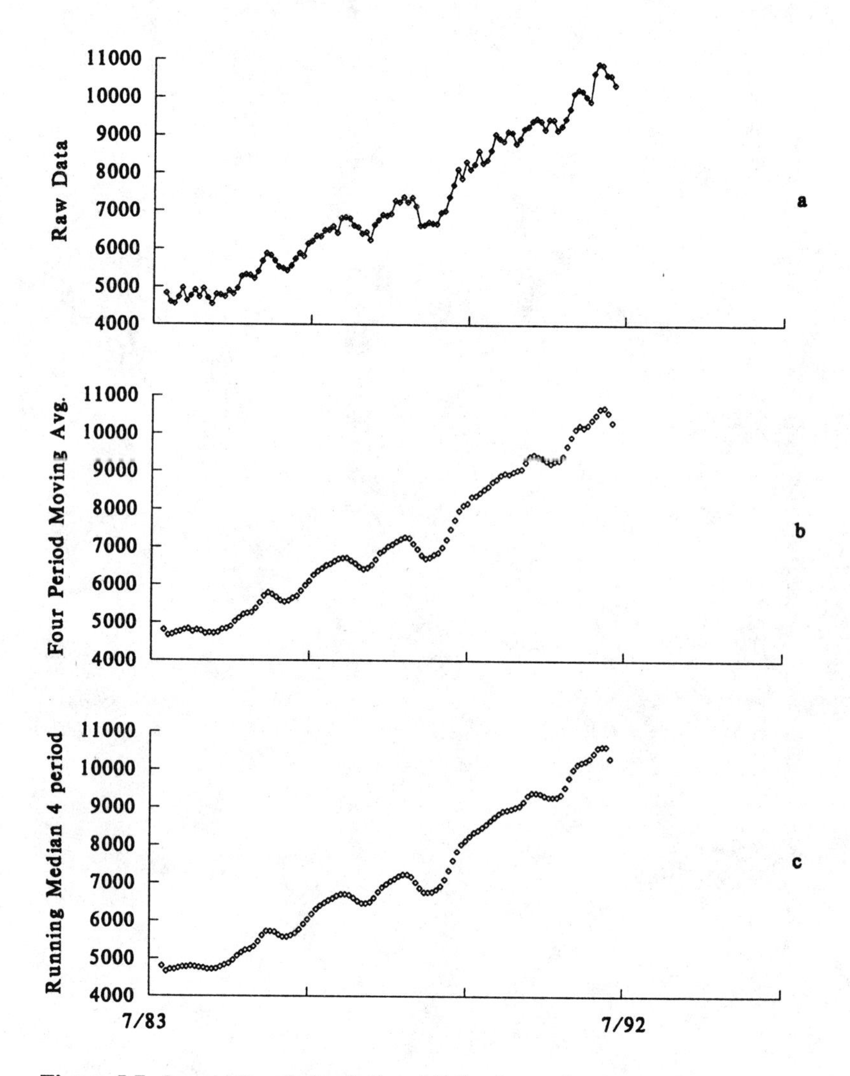

Figure 5.7. Inmate Population in Local Jails: Averaging

The moving average assists in removing some of the random fluctuation of the data, which enables our eyes to follow the overall trend. For data with more fluctuations, a more robust method of taking moving averages can be done by using running medians. The original data are smoothed by a four-period running median in the bottom panel of Figure 5.7. The process is similar to moving averages except that medians are computed rather than means. The curves are smoother than before. This method works especially well with series that have occasional outliers, which tend to distract attention from the images in the data.

These methods use an indirect method for reducing the amount of data—they focus the viewer on the shapes and trends by eliminating some chance fluctuation that makes comprehension more difficult and time consuming. Two other methods directly reduce the amount of data being presented: reducing the number of data points plotted for each year of the series and eliminating some periods in the series entirely. Although both of these methods may be effectively used to communicate data, you must guard against unintentional deception. Omitting part of the data may require an explanation. If the omitted data support a conclusion different from that drawn from the data that are used, the grapher should include the entire series, perhaps adding a second graph that focuses on the period of greatest interest. Loss of credibility for the data and the researcher is too high a price to pay for omitting data.

In Figure 5.8, the series is plotted for a selected month from each year. The graph appears to be emaciated and the smoothed line in the second panel is an improvement. It does work well with the jail capacity data that is added in the third panel. The gap between population and capacity is the number of inmates housed in overcrowded conditions. The bottom panel illustrates a smoothing technique called negative exponential smoothing to make the line connecting the data points smoother than the straight lines drawn between each set of two points in the middle panel. This technique produces an effect somewhat similar to Playfair's trend lines. Of course the difference graph, in this case the inmates in overcrowded conditions graph, could be added here, as it was in the reproduction of Playfair's graph (Figure 5.2a).

Finally, the graphic in Figure 5.9 shows data from the past 4 years to emphasize the recent past. Justification for the use of this portion of the data could be a change in the corrections system, the laws affecting inmate population, or data collection methods. The data do highlight the recent trends, but because there is less of the time series covered in the same size graphic as before, the trend does not appear as steep as it did in earlier series. Is this caused by omitting the data? Not entirely. It is caused by stretching the last 4 years of data over the same distance as the entire series, or in other words, changing the scale. The next section will focus on size and the relationship of the length and width of a graph.

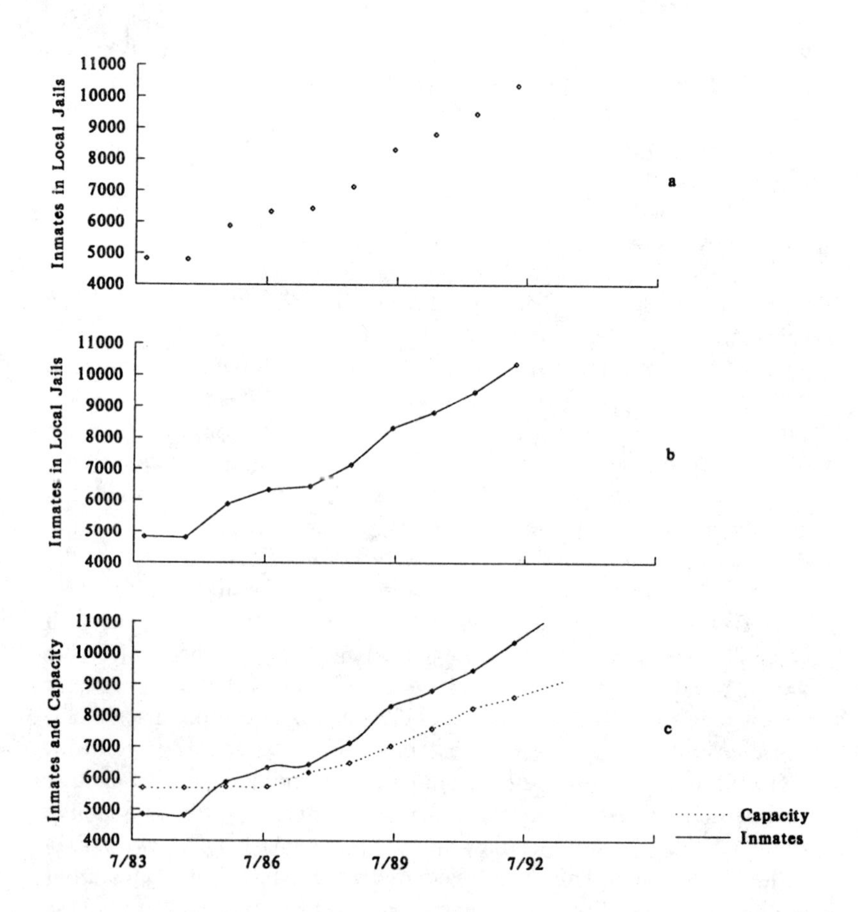

Figure 5.8. Inmate Population in Jails: Yearly Data

SIZING THE GRAPH

Two graphing principles concern the size of a graphic: data density and distortion of effect size. Data density should be maximized (Tufte, 1983). In other words, data should be packed into as small a display space as possible. Of course, human perception places real limits on the size of the marks used to plot the data and how close they can be and still be distinguished. An illustration of the problem of distinguishing differences in graphical symbols was shown in Chapter 4, where limits were placed on the number of units that can be plotted by the size of the

Figure 5.9. Inmate Population in Jails, 1988-1992

page. Along with the limitations on legibility come other limitations such as the quality of common copier reproductions and the aesthetic appeal of the graph in the publication. Dense graphs are preferred, but the degree to which they can be condensed depends on legibility, method of reproduction, and the aesthetic appeal of the graph.

In addition to the overall size of the graph, the relationship between the length and width of the graph, or the scale of the graph, also has implications for its size. Tufte and Cleveland have different ways of approaching the relationship between the length and width of the graph. For Tufte (1983), it is a matter of proportion. He refers to the "golden rectangle" from ancient philosophy, Playfair's work, and some very practical issues about labeling to come to the conclusion that graphs should be wider than tall: "Graphs should tend toward the horizontal" (p. 186). The golden rectangle would have us draw graphs approximately 1.618 times longer than tall. Tufte (1983) seems to set his cap firmly for a ratio between 1.2 times and 2.2 times wider than tall. The two upper panels in Figure 5.10 show these two ratios for the crime rate data in a minimum overall size. Obviously, the choice of scale will affect the viewer's perception of the trend in the data: Panel b makes the crime rate appear to be gradually increasing, whereas Panel a visually depicts a steep increase. Most graph software allows the analyst to choose the length and width of the graph, enabling the graph to be

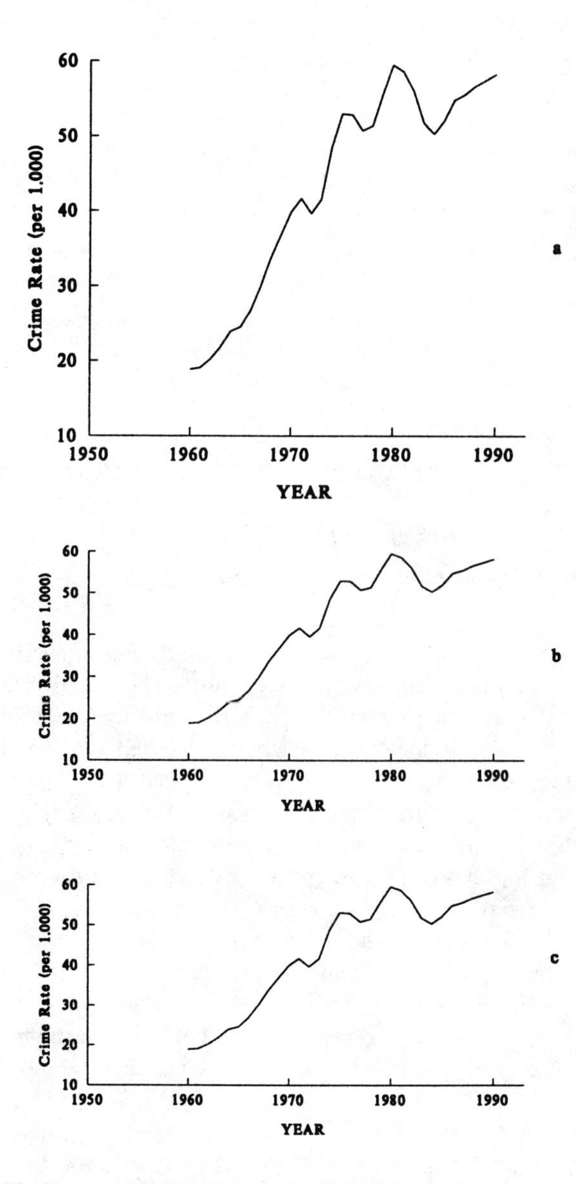

Figure 5.10. Reported Crime Rate in the United States (three scales)
SOURCE: FBI (1991).

sized for the publication space and the appropriate message that is being conveyed.

Cleveland takes another approach to the size issue based upon the slope of the data. He has developed a method for computing the dimensions of the graph to achieve an absolute median slope of 1. The graph plotted in Panel c of Figure 5.10 uses his method and has a ratio of 2.32. Having used this method on several time series graphs, I find it yields a graph at the upper end of Tufte's (1983) recommended ratio, above the golden rectangle. I am aware of one software program, SYSTAT, that will compute Cleveland's (1985) preferred dimensions and size the graph accordingly. For those without this software, a 5-in. wide graph is commonly used in a report on 8½-×-11-in. paper and should be about 2¼ to 2½ inches tall. To follow Tufte's (1983) advice the graph could be shorter, reducing slope. Most software programs seem to default to the square, which gives an impression of urgency.

MINDING THE DETAILS

Ambiguity is the greatest enemy of competent graphical design. Attention to the details of graphics thwarts ambiguity. However, details are those time-consuming nuisances, the minding of which comes after the creative work is done. "I know what it says and they'll figure it out when I explain it," reflects an attitude that may cause a researcher to miss out on an opportunity to communicate data to an important audience. It also misses the fundamental point that Bertin (1967/1983) makes about the power of graphical displays. They are temporal. They can be viewed and reviewed as the interests and curiosity of the audience members peak. To possess atemporality, the graph must stand on its own, as a complete entity.

Completeness involves a number of essential items:

Title: The title should describe the variable(s) and the population being plotted.

Axes: Both axes for the graph should be clearly labeled with the variable name or measure.

Axis lines: Use the minimum number of axis lines. Usually two will suffice, if you need any at all.

Scale: Scales should be marked and labeled at intervals on both axes that permit interpolation of points from the graph for the viewer to retrieve specific data points.

Line labels: Lines should be labeled at the end when the series is multivariate, using the Gestalt principle of continuity; the use of line symbols, such as dashed or dotted lines with a legend, makes reading the graph more difficult.

Source: The source of the data should be noted at the bottom of the graphic for the viewers' convenience.

Grid lines: Grid lines distract the eyes from the data and when placed at arbitrary intervals on a time series do little to aid the retrieval of data; they should be minimized or taken out entirely.

If you review the Playfair graph (Figure 5.2a), you can find a pleasant example of most of these points. The exceptions lie principally in the number of axes and grid lines plotted and where the vertical axis is labeled. Tufte (1983) posits the principle of removing "chart junk" and nondata ink reflected in the statements above. Bertin (1967/1983) emphasizes the need for the title to convey the variable or measure, which he labels the "invariant" to signify that it applies to all the data points. Others have suggested that the title convey a message or the author's conclusion from the graph. I prefer to reveal the data limiting the marks that do not convey data as much as possible, to identify clearly what is being exposed, and to let the viewers come to their own conclusions.

MULTIVARIATE TIME SERIES

Multivariate time series can take three forms:

Cumulative time series
Difference time series
Multiple series time series

The cumulative time series is a set of related time series that are actually parts of a whole. The Medicaid expenditures in Figure 5.3 are one example of a cumulative time series, although they are not plotted as such in this example. Each year of a cumulative time series could be plotted on a bar chart. Difference time series are two time series whose difference is meaningful for analysis. The import-export, balance of trade graphics using Playfair's (1786) data (Figures 5.2a and 5.2b) are one example of this type of time series. The multiple series time series are usually two trends between which the researcher posits a cause-effect relationship. This type of graph usually involves plotting two different metrics on the vertical scale, which makes graphing very tricky. The next three sections describe each type in more detail.

Cumulative Time Series. The cumulative time series is a data series of the components of the whole over time. Although this type of multivariate time series seems easily graphed because the series are in the same metric, it can be quite challenging. Wainer (1992) recounts that an item on the National Assessment of Educational Progress (NAEP) was used to conclude that, "Only 50% of American 17-year olds can identify information in a graph of energy sources" (p. 15) The graph in Figure 5.11 was given to the students. The question asked of them was, "In the year 2000, which energy source is predicted to supply less power than coal?" The graph contains several design errors, including the use of bars for time series data and imperceptible shading differences between natural gas and hydropower sources. Using Wainer's (1992) suggestions to redraw the graph makes the answer obvious, as we see in Figure 5.12. A well-constructed bar chart for the year 2000 predictions would have done as well.

The Medicaid data presented in Figure 5.13 provide an opportunity to review the design choices for cumulative multivariate graphs. First, we must decide if the total is needed. Figure 5.13a shows the five categories of Medicaid expenditures for 1983-1991 graphed as multiple time series without the total included. This graph allows the viewer to follow the individual categories, noting the dramatic rise in inpatient expenditure and the similar pattern in physician and drug expenditures until 1990. In Figure 5.13b, the total is added. The increases in the individual components over time produce a dramatic rise in the total that was not apparent without adding in the total. However, some of the detail for individual categories is lost. The energy source example is in fact a counter example, for which the total was not necessary for the purpose at hand, as illustrated by Wainer's (1992) redesign (see Figures 5.11 and 5.12).

Because of the scale differences between the components and the total, the ability to track individual categories over time is diminished in Figure 5.13b. One solution is to use a log scale to graph the six series, including the total, as depicted in Figure 5.13c. In this graph the log scale values are replaced by the original metric, but the vertical axis is

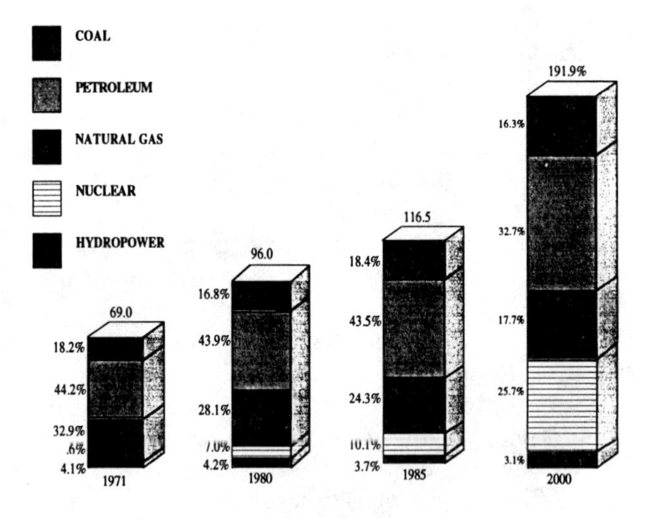

In the year 2000, which energy source is predicted to supply less power than coal?

A Petroleum
B Natural Gas
C Nuclear Power
D Hydropower
E I don't know

Figure 5.11. Estimated U.S. Power Consumption by Source (Quadrillion BTUs)
SOURCE: U.S. Department of Interior, United States Energy Through the Year 2000. Copyright © 1973 Congressional Quarterly, Inc.
NOTE: BTU: Quantity of heat required to raise temperature of one pound of water one degree Fahrenheit.

clearly identified as a log scale. The log scale is useful to those familiar with its properties, but it reduces the slopes significantly, as you can see by comparing the Total series in Figures 5.13b and 5.13c.

Figure 5.13d shows a cumulative graph. This graph appears to be the most pleasing, but it too has its drawbacks. To look at a category of expenditures other than the nursing expenditures at the bottom of the graph, the eye must judge the differences between two curves, a notoriously difficult task (Cleveland & McGill, 1984, p. 549). For this type of graph, it is important to put the category that is the primary focus of attention at the bottom. However, if this category has considerable fluctuation over the time period, it will make the other categories too

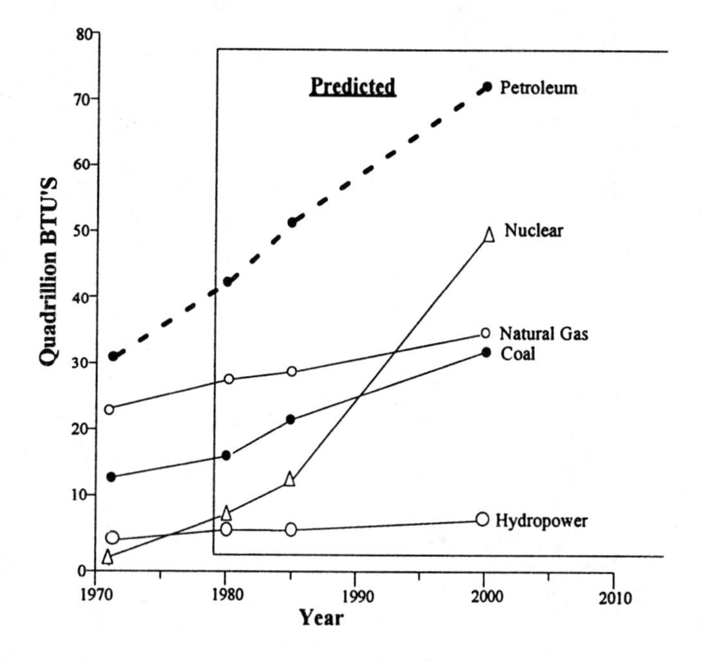

Figure 5.12. Estimated U.S. Power Consumption by Source (Quadrillion BTUs)
SOURCE: U.S. Department of Interior, United States Energy Through the Year 2000. Copyright © 1973 Congressional Quarterly, Inc.
NOTE: BTU: Quantity of heat required to raise temperature of one pound of water one degree Fahrenheit.

difficult to retrieve. In this case, the most stable categories should be placed at the bottom, and the category of interest at the top where the fluctuations can be observed without affecting the perception of the other data.

Difference Time Series. A difference time series involves plotting two series that are not only interesting in themselves but also have an interesting difference. Playfair's chart (Figure 5.2a) showing imports and exports is one example. Originally, Playfair did not draw in the balance of trade; the surplus or deficit was to be ascertained by the viewer based on observing the difference between the two lines. In Figure 5.2b, the balance of trade is added to the graph. Another of our time series can be turned into a difference time series by the addition of another variable: The jail population series can be plotted with the jail capacity to show excess bed space or overcrowding in the jails. In

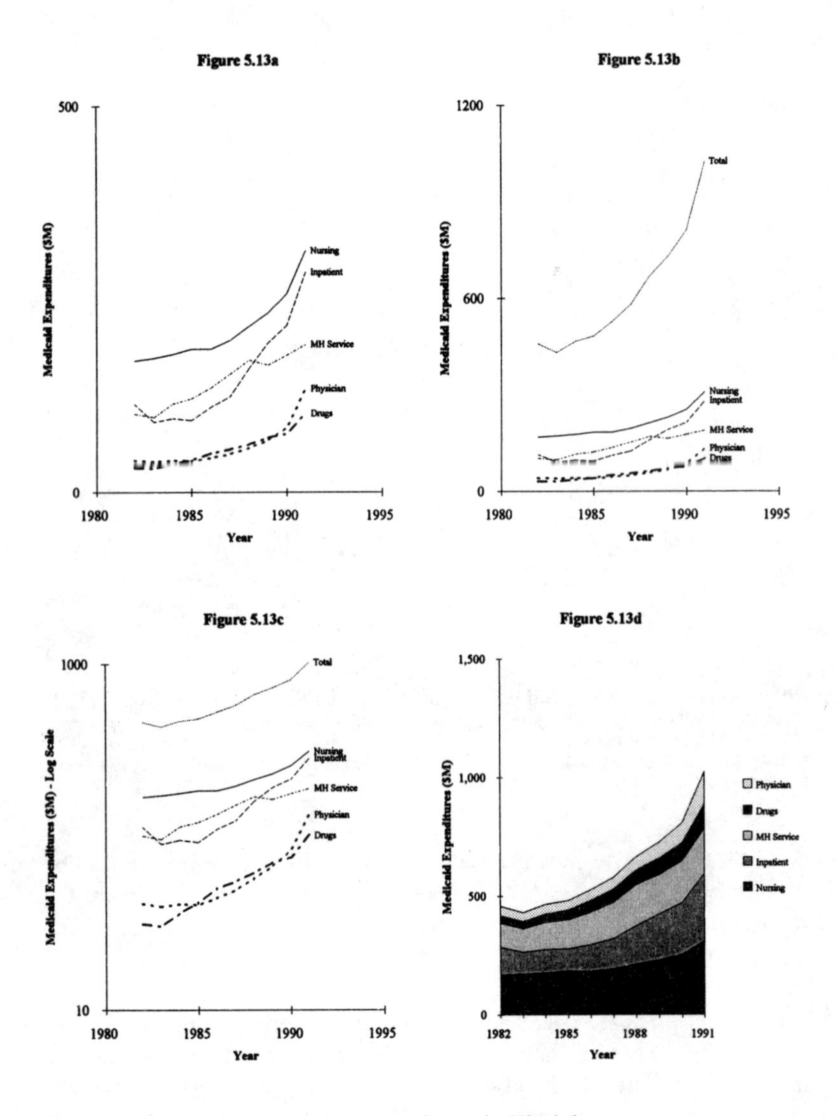

Figure 5.13. Annual Medicaid Expenditures in Virginia

Figure 5.14, the data from Panel c in Figure 5.7 are plotted with a special emphasis on the overcrowding of the jails. Whether to plot the difference as a line is a choice that the researcher must make in the presenta-

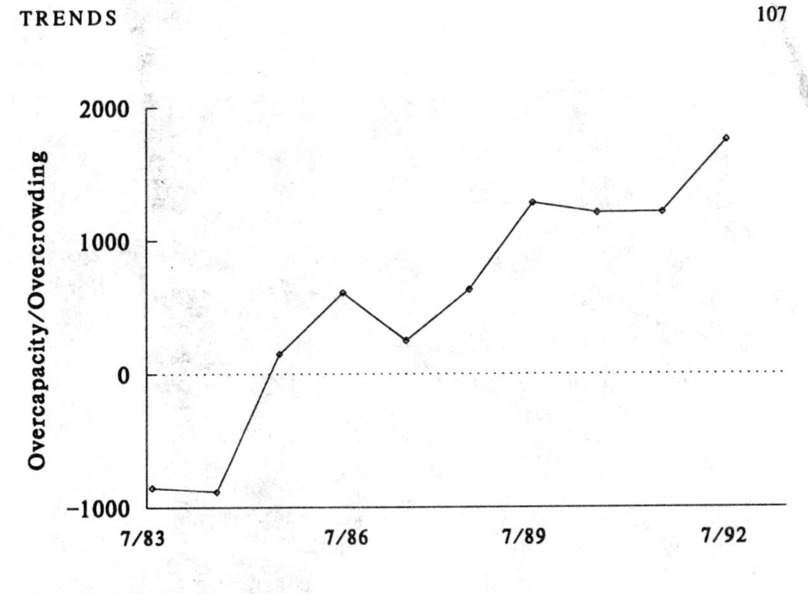

Figure 5.14. Jail Capacity

tion of a difference time series, taking into consideration the impact on scale or the use of a second panel.

Multiple Series Time Series. Multiple series time series are usually an attempt on the part of a researcher to relate dissimilar time series to support or undermine a conjecture of a cause-effect relationship. Wainer (1992) has shown an interesting example of an attempt to use a graphic to support a conjecture that is not supportable by the data (p. 17). *Forbes* magazine published a graph that purported to show the relationship between public school expenditures and SAT scores using a "double *Y* axis" time series plot. The main graph has been reproduced in Figure 5.15. The magazine staff is implanting the idea that educational expenditures have risen faster than SAT scores. Wainer (1992) shows that the opposite conclusion can be suggested by changing the scale of the graph, which has been done in Figure 5.16. In Figure 5.16, SAT scores appear to rise faster than educational expenditures. Obviously we are playing with propaganda in both graphs. Neither conclusion is clearly indicated by the data.

Wainer (1992) eschews the use of this type of multiple series graph. Bertin (1967/1983) suggests indexing both data series from a mean or midpoint and plotting the percent change. In this case I have used 1984 as the index year and plotted the data on a common metric, percent

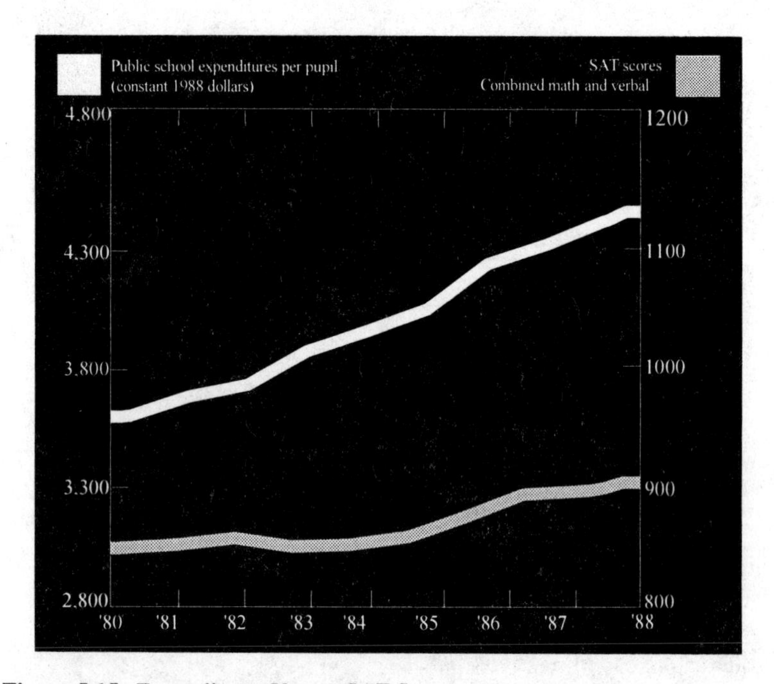

Figure 5.15. Expenditures Versus SAT Scores
SOURCE: U.S. Department of Education: Educational Testing Service, U.S. Department of Labor. This chart is a reproduction of an original graphic found in *Forbes,* May 14, 1990. Reprinted by permission of *Forbes* magazine.

change (Figure 5.17). This allows the graphing of multiple time series without attempting scale manipulation. Of course, the result doesn't allow a causal inference, but it does show the relative trends during the period. The percent change metric reveals a slightly greater percentage increase in educational expenditures than SAT scores over the 9-year period.

CONCLUSION

Time series graphs represent a significant step in the evolution of graphical displays. They were the first graphs that did not rely on spatial analogies as had their predecessors—maps and diagrams. They set the stage for two-variable, or Cartesian plots, which are the most abstract and first truly relational graphs. Time graphed from left to right represents

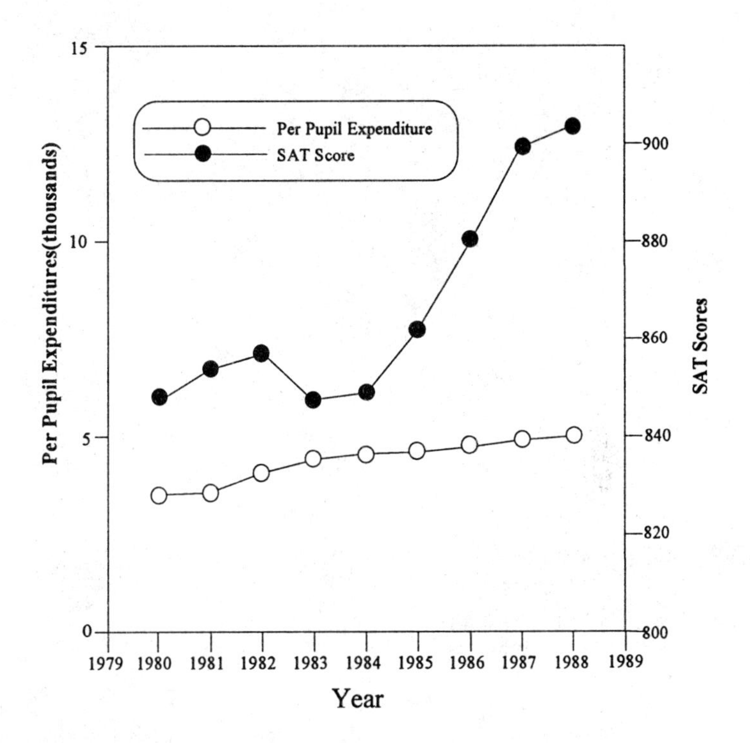

Figure 5.16. Expenditures Versus SAT Scores
SOURCE: U.S. Department of Education, Educational Testing Service; U.S. Department of Labor.

an analogy to a dimension in the physical world. In fact, the longstanding popularity of time series graphs may stem in part from our ability to easily grasp a graph with an analogy to the physical dimension of time. Clearly, the next level of abstractness in graphics, the two-variable plot, has not received near the level of usage, even though it is more versatile and can present more complex data.

In the next chapter, the two-variable plot will be described and displayed. The two types of displays share many similar properties, as the time can be regarded as a particular variable graphed on the horizontal axis. Concerns for shape, scale, and size will carry over to the general case of the two-variable plot. Also, the concern for details, such as titles, labels, axes, and legends will be relevant. Many of the smoothing techniques will be applied to those graphs, although they will become norms rather than trends and the variations around the norms

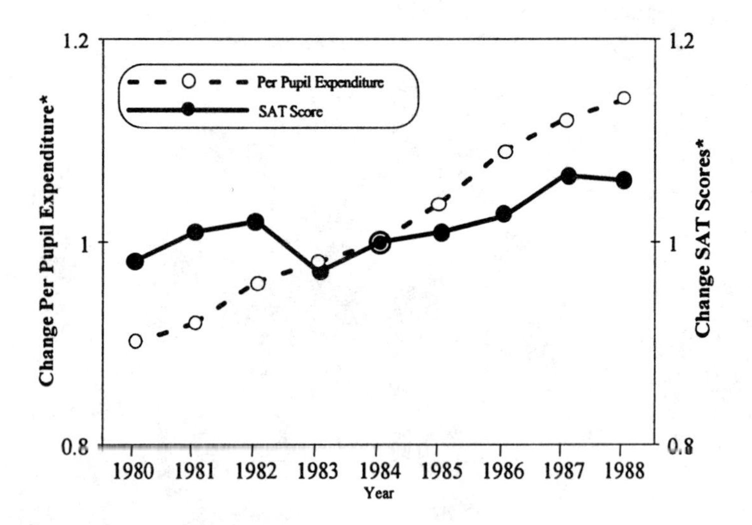

Figure 5.17. Expenditures Versus SAT Scores
*Both scales were indexed using the 1984 figures for each.
Per pupil expenditure = $4000
SAT score = 851 (combined math and verbal)
SOURCE: U.S. Department of Education, Educational Testing Service; U.S. Department of Labor.

will take on a different meaning. Questions about the fluctuation around trend lines typically go straight to the time periods that appear to be exceptions to the trend. We start by asking, "Why?" For two-variable plots, we must first ask, "How well does the line or curve summarize the data?" Once we are satisfied that the line is a reasonable norm, we can go on to questions of cases that appear to be exceptions. It is the fact that the mark or data point on the two-variable plot represents a case that is not concretely identified that sets it apart from the time series, in which every point is identified by a unique time period. For each series there is only one point that corresponds to the time period on the horizontal axis. In a two-variable plot there may be several cases identified with a particular value on the horizontal axis, and no cases identified with others.

EXERCISES

1. Review the graphs that you collected for the exercises in Chapter 1 or collect a new batch. What percentage of the graphs are time series? Classify them as univariate, cumulative, difference, and multiple. Analyze each of the

designs. Consider how much data is plotted. Can you extract information at the three levels of graphicacy from the graphs?

2. Select a graph representing each type and redesign the graph as needed. Make sure you add the labels, scale indications, and other details.

3. Using a recent volume of the U.S. Census Bureau's *Statistical Abstracts* (1993) or another source of time series data related to your field, plot a simple time series and a multivariate time series. Explore the visual implications of changing the scale of the univariate graph and adding smoothing lines. Prepare at least two alternative designs for the multivariate graph. Before designing the alternatives, decide what the graph is intended to convey.

6

Graphical Complements for Correlation and Regression

Applied researchers and evaluators commonly use correlation and regression to analyze relationships between two or more variables. From a statistical viewpoint these techniques have common traits and fall under the umbrella of linear models. Not surprisingly, their graphical complements also are related. Graphical displays for both techniques involve relational graphs of two or more variables. Relational graphics are plots in which the symbol system does not directly represent any physical characteristics: The variables are plotted in relation to one another. These graphs are often referred to as Cartesian plots, *X*-*Y* plots, scatterplots, and 2-D or 3-D graphs, depending on the number of variables plotted.

Graphing relationships between two variables, neither of which is space or time, represents the most abstract use of the two planes of the display. In these graphs, the full potential of two-dimensional graphical displays are realized. The relationships appear without physical analogies. The eye is able to find patterns and exceptions to patterns. Elementary graphicacy (Level 1) can be extracted for both variables; norms and more complex comparisons can be obtained with relative ease, especially for those familiar with the format of the Cartesian plot.

The abstract nature of these graphs opens up many possibilities for using them. Neither time series or physical coordinate data are needed. Any data set with multiple cases or observations and two or more variables can be plotted with these relational graphs. The versatility of these plots combined with their capacity to convey information accounts for their popularity in communications between applied researchers. The format has been learned (stored in long-term memory) through frequent encounters in textbooks and journal articles. Graduate education installs the Cartesian plot in the long-term memory of most applied researchers, but we cannot assume that all members of an audience will readily comprehend these abstract plots. For applied researchers to realize the potential of these graphs, audience members must retain the relational graphic form in long-term memory through

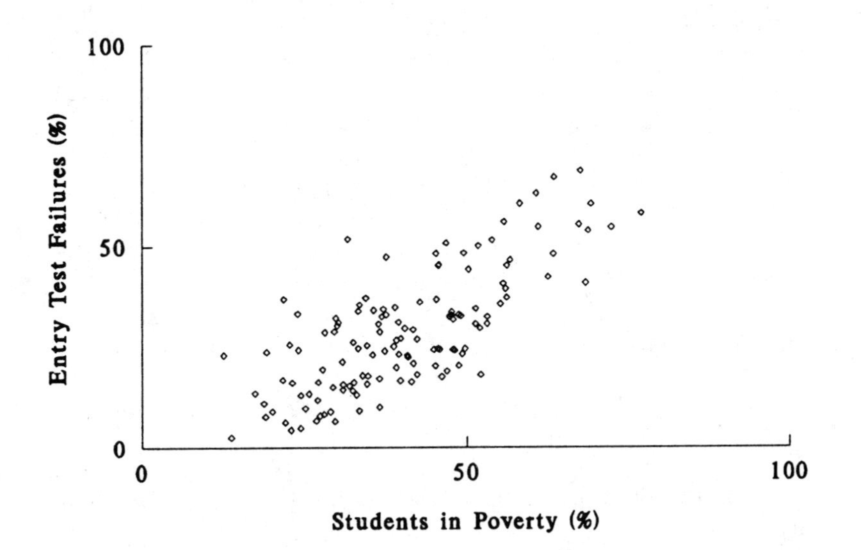

Figure 6.1. Poverty and Entry Test Failures in Virginia School Districts

frequent exposure and be motivated to retrieve information from a particular graph.

The abstract nature of Cartesian plots quite possibly accounts for their lack of use in popular publications (Tufte, 1983) and in briefings and reports of applied researchers for lay audiences. Relational graphics are complex because they are abstract. Because two-variable plots are only rarely addressed to applied research audiences, data tables and narrative descriptions of bivariate relationships appear in their place. The power of the bivariate plot to inform is shown in Figure 6.1. The graph shows the relationship between student poverty and school readiness in Virginia school districts. Without much explanation, we see that in a school district where more of the students live below the poverty level, more students are likely to fail at least one part of their school readiness tests. When the poverty level reaches 60% in a district it appears likely that about 50% of the students will fail an entry test exam. For districts with 25% poverty, the expected percentage of tests failures appears to be about 10%.

The pattern appears to be tightly correlated and linear. The Pearson correlation coefficient for these variables is .75. Many who would not grasp the significance of this statistic can grasp the significance of the relationship from the graph.

The relationship is hard to convey in words as succinctly as it is communicated in the graph. The Pearson's r of .75 will not make an impression on an audience whose members are unfamiliar with statistical terms. In fact, a correlation at this level says little about the nature of the relationship to anyone, as Cleveland et al. (1983) have shown (pp. 78-79). To state the finding that the percentage of the students failing one part of the readiness test given at school entry is directly related to the percentage of the poverty students is a strong statement. But it doesn't have the power of the graph: The graph convinces us through our visual sense, not our intellect. Further, the process of coming to their own conclusions may strengthen the researcher's and the viewer's belief in the conclusion. An inescapable conclusion from the graph is that the educational task will be more difficult in the high-poverty districts.

Relational graphs complement some of the most popular statistical methods. Correlation and regression graphs use two variable plots. Both the X and Y plane are defined by a continuous variable in these graphs. In the case of regression, a third dimension is often useful, specifically, adding a third continuous variable by varying the size of the plot symbol proportionately to the value of this variable or adding a categorical variable by using the plot symbol to identify a group or groups.

In the next section of this chapter, I examine some of the graphing possibilities for bivariate correlation. Then I discuss two multiple-variable correlation graphs, one of which is extremely useful in analyzing the results of cluster analysis. Regression, the next topic, has many graphical complements, some descriptive, some analytical, and some diagnostic. For the regression complements, we will explore various forms of smoothing as well as graphing residuals (difference between the regression estimates for an observation and its actual value). Finally, we will deal with multivariate regression.

CORRELATION

The correlation between two variables is generally represented graphically on a Cartesian plot. Because the nature of correlation analysis is symmetrical, the plot is an attempt to show whether or not the data fit a discernible, compact pattern. Scale becomes very important in the perception of association (Cleveland, Diaconis, & McGill, 1982). The abstract form of the graphic means that visual perception of pattern is

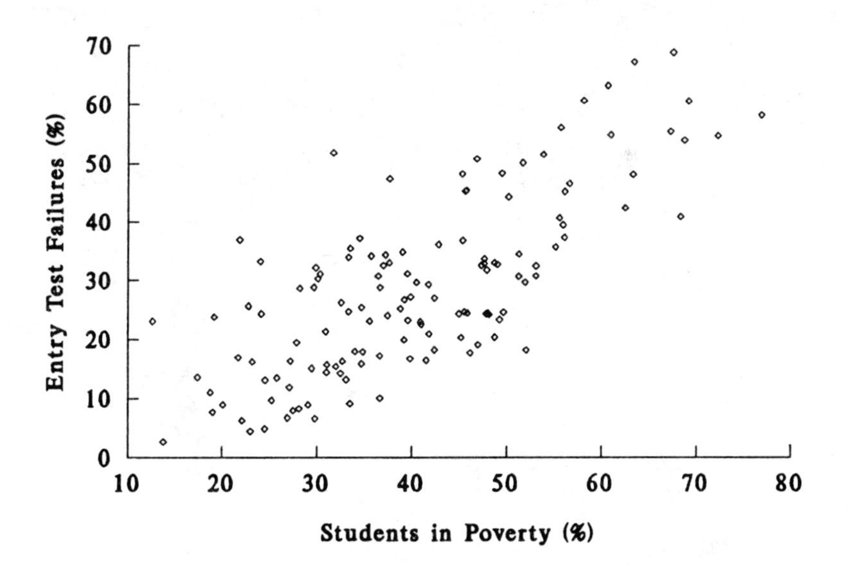

Figure 6.2. Poverty and Entry Test Failures in Virginia School Districts

relative to the data points as they are distributed over the plane of the paper. There is no external indication of the norm in this type of relational graph. Figure 6.2 again shows the correlation between the percentage of the student body living in poverty and the percentage of students that fail one or more parts of an entry test for school divisions in Virginia.

This is a second plot of the data plotted in Figure 6.1, changing only the scale of the plot. Does this plot appear to be more or less correlated than the previous plot? Cleveland et al. (1982) found that the plot in Figure 6.2 is generally perceived as less correlated, though both plots are of the same data. Compacting the scale increases the perception of the amount of correlation. This should underscore the importance of scale in relational graphs. Generally, bivariate plots should fill the space defined by the axes, and the width should be 1.2 to 2.2 times the height. The scale and beginning and endpoints on each axis should be arranged so that (a) no data points are omitted from the graph, (b) the slope approximates a 45 degree angle, (c) there are data points plotted in a frame just inside the four axis lines (if all four were drawn) and defined by the lowest and highest sets of tick marks on each axis, and (d) no data points are plotted directly on the axis lines.

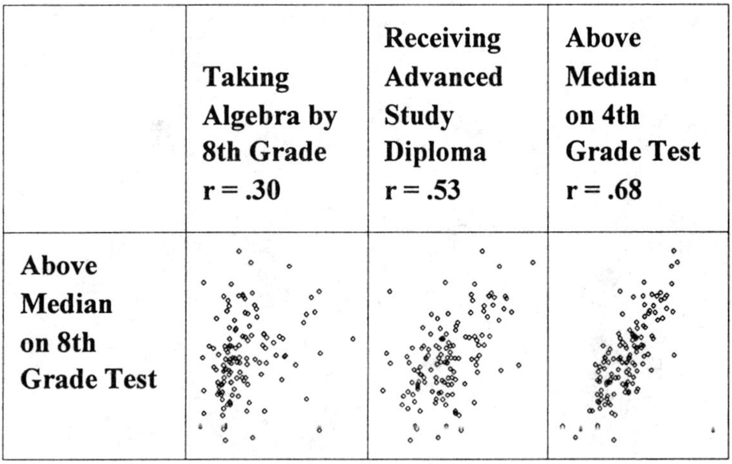

	Taking Algebra by 8th Grade r = .30	Receiving Advanced Study Diploma r = .53	Above Median on 4th Grade Test r = .68
Above Median on 8th Grade Test			

Figure 6.3. Three Correlations With Percent Above Median Test Scores

Bivariate correlation plots show patterns, but they also encourage looking at exceptions or variations. Two school districts that are exceptions to the general pattern can be seen in Figure 6.2. One lies at about 52% poverty and the other at about 68% on the poverty scale. Their percentage of entry test failures are about 18% and 40%, respectively. The audience can also pick out three school districts that have higher levels of failure on the entry tests than might be expected, given their poverty level. Looking at the data allows the audience to grasp the complexity that the researcher understands: Although the relationship between living in poverty and coming to school relatively unprepared is strong, it is not deterministic. There are exceptions. The pattern is not countered by the exceptions, but a more complex state of affairs is discernible to the audience. Outliers or exceptions to the pattern do occur and understanding the factors that may have caused the exceptions is likely to be an important next step.

In Figure 6.3, three correlations are shown. The correlations increase from .30 to .53 to .68. These graphs provide perspective on the correspondence between the level of correlation and the visual perception of correlation in a bivariate plot. These three plots are from the same batch of data as the earlier two figures. The correlation between the percentage of students taking algebra in the eighth grade and the percentage

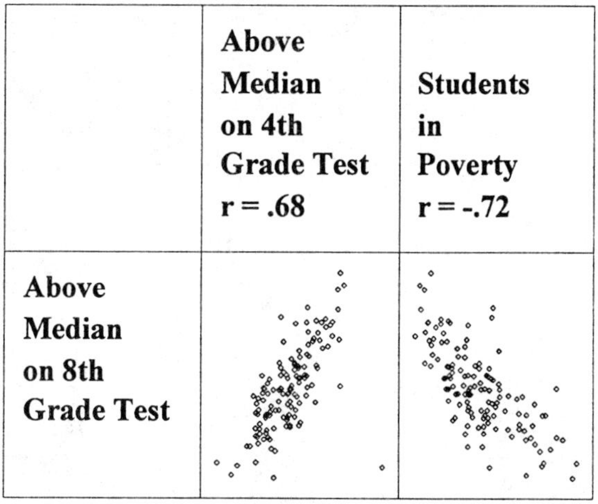

	Above Median on 4th Grade Test r = .68	**Students in Poverty r = -.72**
Above Median on 8th Grade Test		

Figure 6.4. Positive and Negative Correlations

scoring above the statewide median on a standardized test shows a modest correlation of .30. Those scoring above the median are more closely correlated with the percent receiving an advanced studies diploma (r = .53). The correlation between the percent above the median in eighth grade and percent above the median in fourth grade is the strongest (r = .68). Moving from left to right, the scatter of points in each panel takes a more concentrated shape. The plot on the left has little discernible pattern and several points spread wide to the right of the box. The plot on the right illustrates a tightly packed set of points with one outlier on the lower right side of the box. If it were not for this outlier the correlation would be much higher.

Because correlations run from −1 to +1, the contrast between positive and negative correlations is also useful to illustrate. The first box in Figure 6.4 repeats the +.68 correlation shown in the previous figure. Next to it the plot of percentage of students scoring over the state median in eighth grade and the percentage of students in the district eligible for free and reduced-price lunch is shown. The Pearson's correlation coefficient for these variables is −.72. The correlation is slightly higher than in the left-hand box, but there appears to be more scatter and a slightly less well defined pattern that runs at a diagonal to the companion plot. This is an excellent illustration of the impact of an

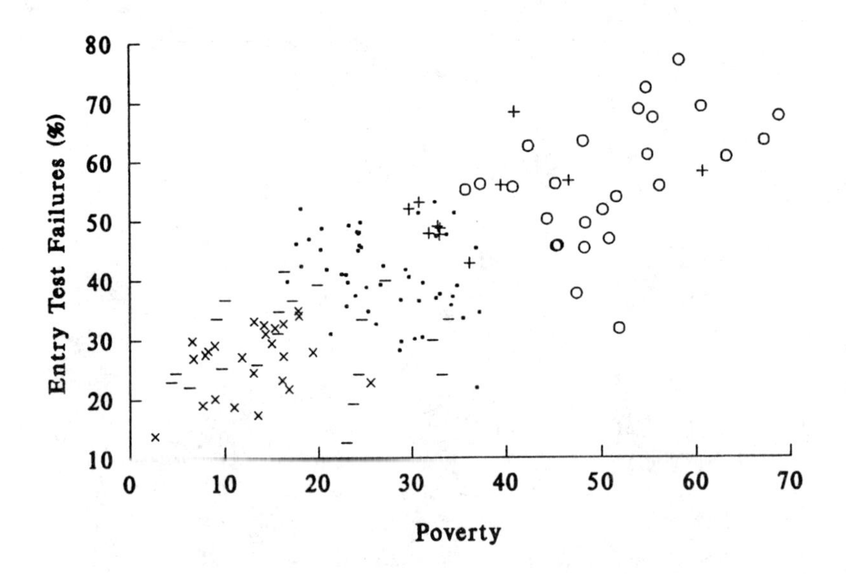

Figure 6.5. Cluster Groupings

outlier, such as the one in the lower right corner of the left-hand box, on the correlation coefficient. Without the visual inspection of the relationship between these variables, the outlier would have gone undetected and a researcher may have assumed that the relationships would be virtually mirror images, as one relationship is positive and the other negative.

Cluster Analysis. Bivariate plots, which are similar to these correlational plots, can be used to examine the results of cluster analyses. Cluster analysis is similar to correlation analysis in that the relationships between the variables are symmetric. That is, these types of analysis are used to examine the relationships between two or more variables without assuming that one variable affects or "causes" another. Regression, on the other hand, is an asymmetric analysis, where one variable is expected to be influenced by other variables.

For cluster analysis, it is often important to understand the threshold effects or how distinct the clusters are for each variable that is used in the analysis. Cluster designations can be shown by plot symbols to enable viewers to discern the amount of mixing of closely located cases among different clusters. Figure 6.5 shows one such plot using the same Virginia school data and the clusters formed from a four-variable clustering procedure. Four variables—poverty measured as above by percent of students eligible for free lunch, average income, a wealth index, and school entry test failures—were used to form five clusters of school districts. The five symbols in the figure were selected to maximize their visual distinctiveness. Bertin (1967/1983) has suggested that the three shapes for symbols that present the greatest visual distinctiveness are the point, the dash, and the cross or x (p. 178). To those, I added the circle and the plus. The plus relies on angles to distinguish it from the cross; the circle sufficiently varies in size from the point for an easily perceptible distinction.

Points, which represent the first cluster, dominate the center of the plot in Figure 6.5. The circles, representing Cluster 2, are spread across the upper right quadrant of the graph, and Clusters 3 (dash) and 5 (x) are mixed in the lower right corner. The few members of Cluster 4 (+) are huddled about the boundary of Clusters 1 and 2. Clusters 3 and 5 seem to be school districts with few students from poverty homes, where children report to the first grade ready for class. Cluster 2 seems to comprise the disadvantaged districts that deal with large numbers of students who are behind on the day they enter the first grade. From this diagram we have a sense that Cluster 2 (o) is as distinct from Cluster 1 as it is from Clusters 3 and 5. But we might question the distinctiveness of Clusters 3 and 5. Because Figure 6.5 presents only two of the four variables used to form the clusters, we should expect that other variables may account for the distinction between Clusters 3 and 5, as well as helping to characterize Cluster 4.

The next graph, Figure 6.6, is particularly useful in extending the two-variable exploration that was described above to displaying differences and similarities that occur across all four of the variables used to form the clusters. Each box in the graph is a bivariate plot of the two variables that define the column and row; in other words, the plot is analogous to a correlation matrix. The bivariate plot in Figure 6.5 lies in the lower left corner of the matrix in Figure 6.6 and uses the same plot symbols as described for the bivariate plot. We can begin to grasp the distinctiveness of Clusters 3 and 5 by looking at the column identified by the wealth index variable and locating the x and dash symbols. The third cluster (dash) is characterized by high wealth index values, which distinguish these districts from Cluster 3. Clusters 3 and 5 are distinct in the window that shows the plot of entry test failures by wealth index. Cluster 4 (+) is also distinct in that plot, appearing to lie in the quadrant high on the wealth index values and in the upper range on the percentage of entry test failures (the upper right quadrant).

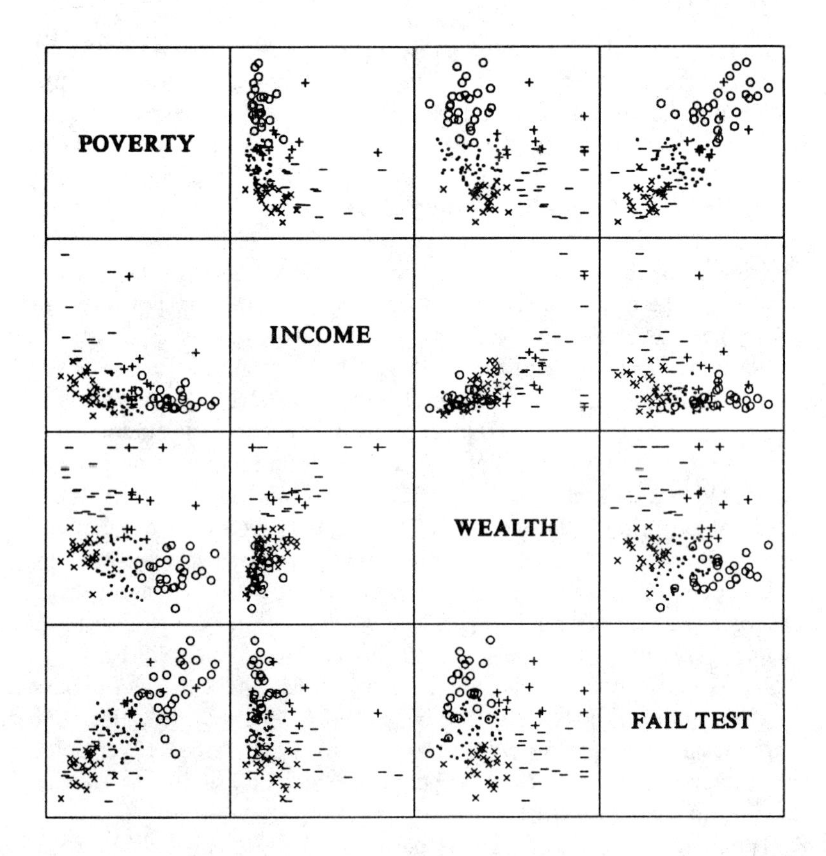

Figure 6.6. Plot Matrix for Cluster Analysis

As expected from a multivariate technique such as clustering, a multivariate graphical display provides the greatest insight. The clusters appear most distinct in the two plots of the wealth index and percent poverty. But visual analysis of the differences allows the analyst a clearer perspective about unique characteristics of each cluster and can be useful in labeling and describing the clusters to the audience. Obviously, the perceptible distinctiveness of the cluster symbols diminishes as the size of the graphs diminishes. Thus there is a lower limit, which Bertin (1967/1983) finds to be at about 2 mm, below which legibility is lost (p. 178). Without the symbols for cluster designations, the plot matrix is a useful supplement for viewing the relationships that are summarized by correlation coefficients.

The use of symbols as the plotting marks is a clever way of introducing a third variable into a two-dimensional scatterplot. Symbols can be used for a variety of categorical variables, such as cluster identifiers, region, or treatment group membership. Using variable plot symbols also allows analysts to make specific units distinct, such as those selected for in-depth case studies or on which they wish to focus attention. A graph such as this can be quickly customized for a particular audience that the researcher knows has an interest in a particular unit or units. Experience shows that there are practical limits to the number of groups identified with plot symbols. Two groups are easiest and requires little time for recalling the symbol codes; five is about the upper limit of distinctions. Many graphical software packages allow the researcher to use a variable in the data set for identifying the plot symbols. Usually this requires either transforming or adding the categorical variable to the data set so that the values for the new categorical variable correspond to the numeric codes for symbols that are listed in the software instruction manual. Remember the five symbols used in Figures 6.5 and 6.6 to create the greatest visual distinction for the graphical display: A table with the symbols and their corresponding code numbers taped to the computer monitor can save time in selecting the codes.

Matrix plots of the sort shown in Figure 6.6 are referred to as "draftsman displays " in Chambers et al. (1983). These can be especially useful in the selection of independent variables for regression analysis and the detection of multicollinearity. In SYGRAPH they are referred to as "scatter plot matrix graphs," or SPLOMs. As the example shows, these plots are not useful for reading data values directly from the graph. Figure 6.6 does not show scale values. The plots support higher order graphicacy functions, as was true for the multivariate graphs in Chapter 4. These graphs give a sense of the patterns and norms in the relationships and allow multiple comparisons from a single graph. As in the previous chapters, trade-offs in what a graph attempts to convey must be made. In this case, sacrificing descriptive functions for analytical ones yields benefits that cannot be garnered if both are attempted. Now we will turn to regression and some of its graphical supplements.

REGRESSION

The most simple and most useful regression plots for use with the audience for research results are identical to correlation plots in that the

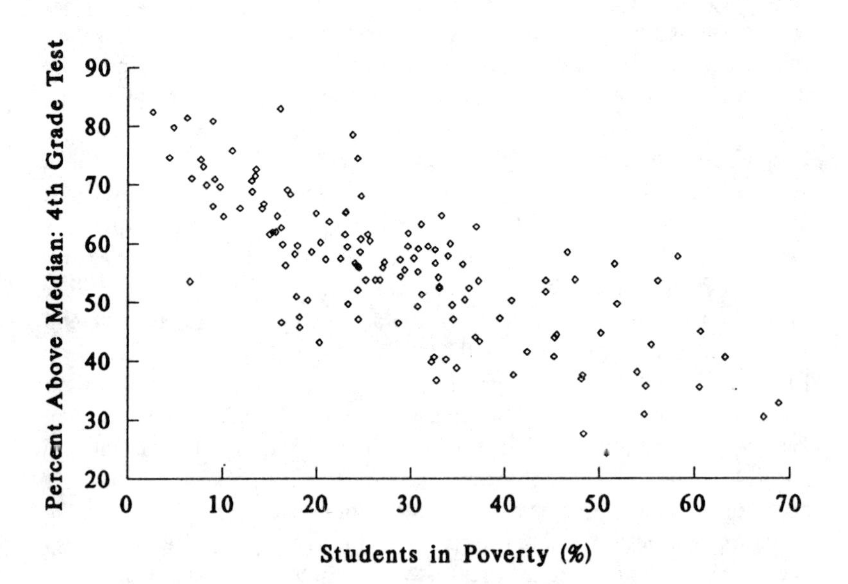

Figure 6.7. Poverty and Test Scores in Virginia School Districts

points are plotted as a relational graph. Figure 6.7 shows an example of a relevant plot from the same data that I have used in the earlier graphs. For this example, we are examining the relationship between the level of poverty as measured by the percent of students eligible for free and reduced-price lunch and the percentage of students who score above the state median on a standardized test in the fourth grade for school districts in Virginia. In this graph, the convention is followed in graphing the "predictor" or hypothesized causal variable on the horizontal (X) axis and the response or effect variable on the vertical (Y) axis. The result, as we would expect, shows that the relationship between poverty and test performance is negative and appears to be both strong and linear: more poverty in a school district results in lower test scores.

The asymmetric nature of the relationship under examination allows some useful additions to regression graphics. Linear regression, the most utilized regression technique, results in a line that minimizes the squared vertical distances from the line. The line that summarizes Y in terms of X has been added in Figure 6.8. Overall, the line appears to summarize the data well, though the line runs below most of the scores in the lowest range of poverty. Although regression produces the best

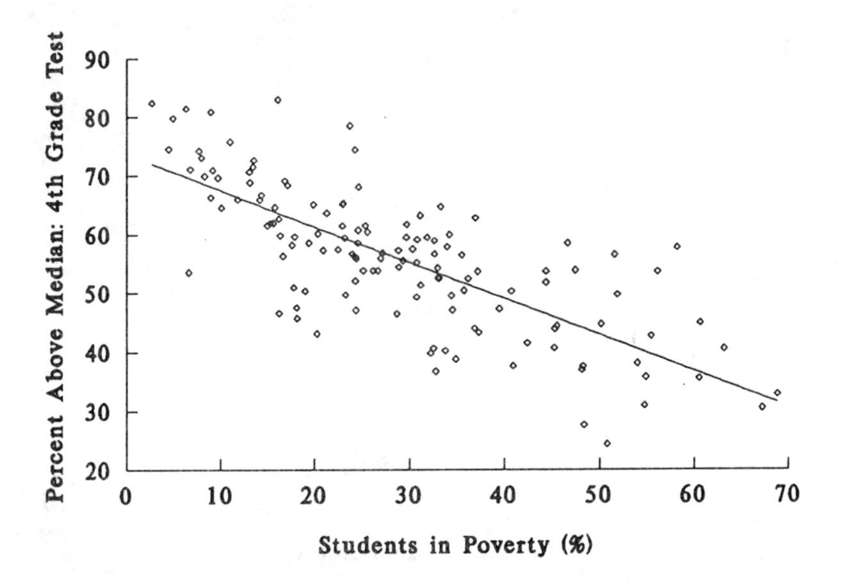

Figure 6.8. Poverty and Test Scores in Virginia School Districts: The Regression Line

linear, unbiased estimates to compute the line, it can be greatly influenced by observations with extreme values. No extreme outliers are obvious from this graph, but we will return to this issue in later graphs. In addition, a line may not summarize a relationship well if it is not linear. This relationship does appear to be linear.

Figure 6.9 uses a technique developed by Cleveland (1985) known as robust locally weighted regression (LOWESS) to examine the linearity of the a bivariate relationship. LOWESS calculates the middle of the distribution of y for every value of x. For each value of the horizontal variable, its nearest neighbors are selected, and weighted least squares are used to fit a line. The value of the line at that x represents a point on the LOWESS line. LOWESS uses a parameter that represents the proportion of points used for the selection of nearest neighbors; for this case .8 was chosen. The curve at the left side of the LOWESS line highlights a problem at the lower range of percent of students in poverty, but generally seems to reinforce the earlier observation that the relationship is sufficiently linear to choose a linear regression approach.

Another graphical display that complements regression analysis is the plotting of residuals, or the differences between the actual and

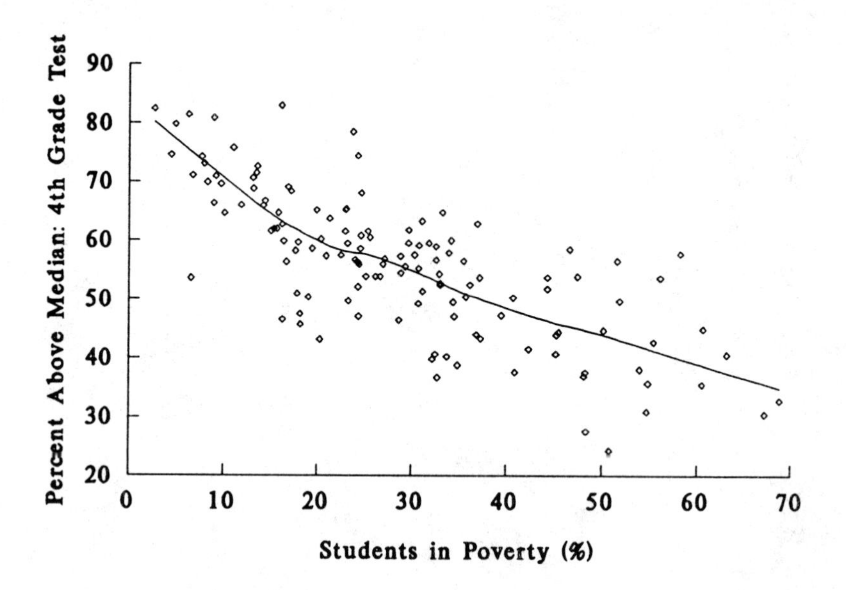

Figure 6.9. Poverty and Test Scores in Virginia School Districts: LOWESS Smoothing

predicted values for each observation. Having satisfied ourselves that the relationship is reasonably linear, we use least squares to compute the equation:

$$Y = 73.6 - .612X$$

$$R^2 = .56$$

The coefficient is significant at the .05 level. Figure 6.10 shows the plot of the residuals for this equation by the independent variable, percent poverty. The horizontal line is plotted at $y - \hat{y}$ and creates a smoothed line from which to judge any pattern in deviations. This plot helps to identify unequal error variances, which can bias the coefficients, as well as identify outliers. The points appear to be scattered around the residual = 0 line, except in the low range for the poverty variable, once again pointing out the problem mentioned above. The plot of residuals is principally an analytical and diagnostic tool for the analyst. It does allow the audience to grasp the concept of probabilistic

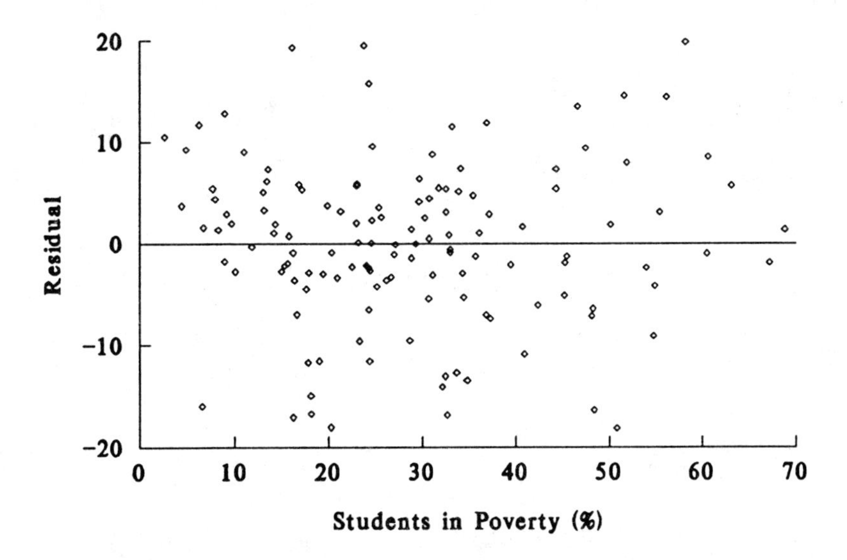

Figure 6.10. Poverty and Test Scores in Virginia School Districts: Residual Plot

rather than deterministic relationships and to avoid the conclusion that poverty completely explains the results on standardized tests. Categorical variables can be used as plotting symbols in this graph to focus on specific units, if useful for the analyst or the audience.

Another common question in regression analysis involves the decision to add another explanatory variable. In this case, we may wish to consider adding the percentage of students that failed an entry test to improve the explanatory power of the model. Panel a in Figure 6.11 shows the bivariate relationship between entry test failures and the fourth-grade test score measure. The display in Panel b shows the LOWESS curve and provides some confidence that the relationship is approximately linear. In both figures, note the outlier at the far right of the display.

To provide a comparison for a nonlinear relationship, Figure 6.12 shows the relationship between the percentage of college graduates in the community and the percentage of fourth graders who scored above the statewide median. A transformation in the percentage of college graduate variable to attempt to make the relationship linear before least squares might be appropriate in this case.

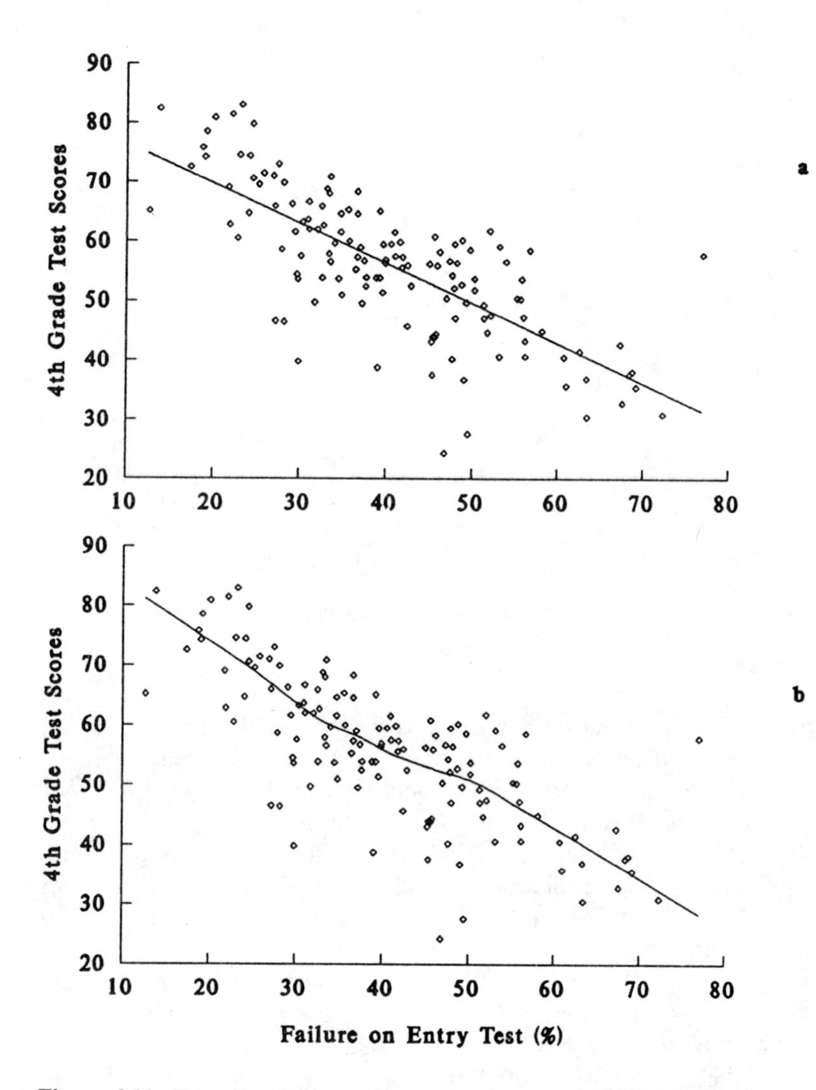

Figure 6.11. Entry Test Failures: Regression Line and LOWESS Smoothing

To investigate the possibility of the regression with two independent variables further, a three-variable plot can be used. Figure 6.13 shows the plot that was used earlier with the size of the symbols representing the percentage of students failing an entry test for the observation. The dots grow larger toward the lower right corner of the graph, indicating

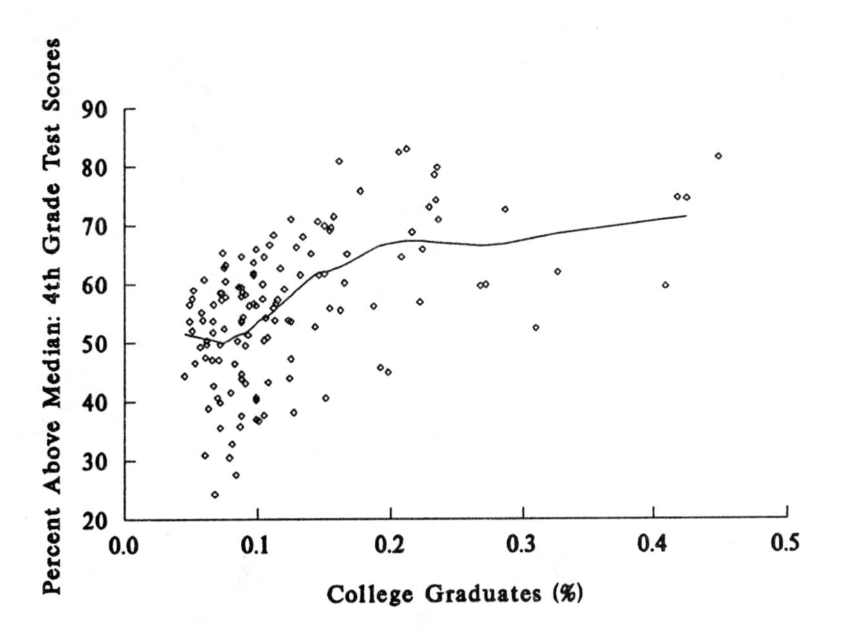

Figure 6.12. LOWESS Smoothing

a positive correlation with free lunch eligibility and a negative correlation with test score performance, as expected. The fact that the dots are larger on the lower half of the graph indicates a relationship between the test score performance and students failing an entry test. Larger dots below the regression line indicate that this relationship holds even after taking free lunch eligibility into account. The dots do appear to be somewhat larger. A few statistical software programs will allow an analyst to use a variable to define the size of the plot symbols. SYGRAPH was used for these plots. A ratio of 1:7 of the lowest to highest values for the variable being used for size produces the optimal visual distinctiveness. Sometimes this requires transformation of the data, which should be done in a monotonic fashion to preserve the original distinctions. However, recalling the usual bias in the perception of area differences—typically they are underestimated—may lead the analyst to exaggerate the differences by using a power function in the data transformation.

Another way to look at the utility of adding an additional variable into the equation is shown in Figure 6.14. This figure plots the residuals that result from regressing the original independent variable on both the

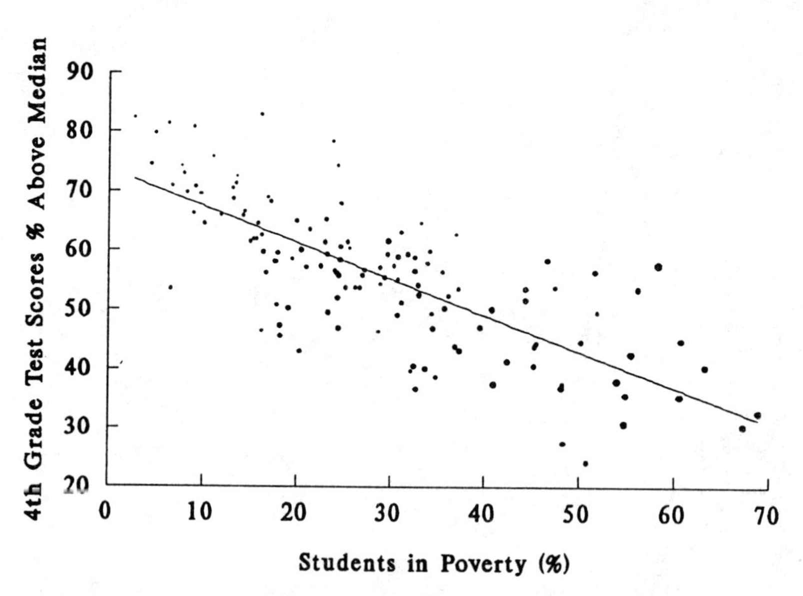

Figure 6.13. Poverty and 4th-Grade Test Scores: Size = Entry Test Failures

dependent variable and the proposed additional independent variable. The regression essentially partials out the effect of the first explanatory variable and allows a visual examination of the added impact of the additional variable. The display shows a modest effect, but one that should be tested further. When both variables are added to the right side of the equation, $R^2 = .63$, and both coefficients are significant. Despite the correlation between the two independent variables ($r = .75$) that was displayed in Figure 6.1, the impact of multicollinearity does not appear to be a problem because both coefficients are significant, the signs are both negative, as expected, and the equation gains in explanatory power when both are included.

SUMMARY

Relational graphs, which are used to complement correlation and regression analysis, are the most abstract graphical forms. They are also some of the most useful in the applied sciences. Ironically, their abstract nature makes them some of the least utilized types of graphical displays

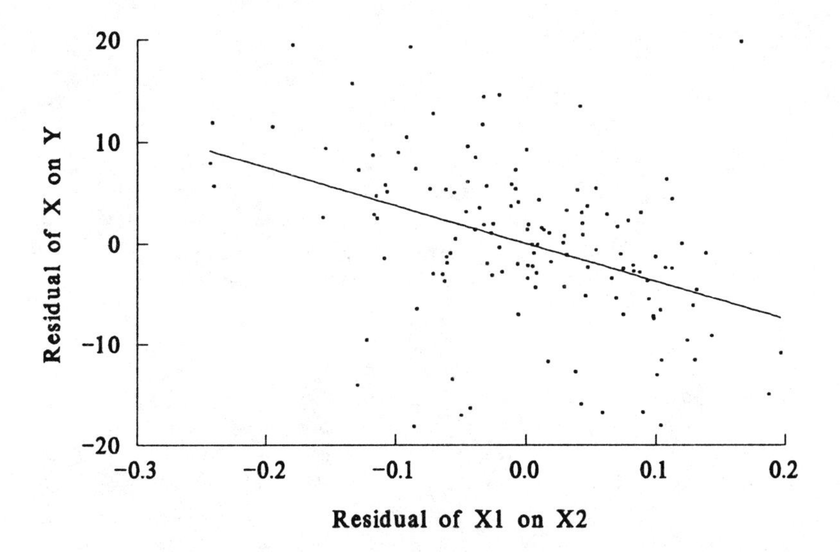

Figure 6.14. Residual of *X, Y* by Residual of *X*1, *X*2: Regression Line

directed to audiences for applied research results. The inability of the audiences for applied social research to understand simple scatterplots can be overcome only by using them. This is the paradox that we face in communicating more information to individuals outside our areas of substantive expertise. Many viewers are not comfortable with decoding the information in relational graphs. However, only if we provide them with more informative graphs will they begin to develop the graphical decoding skills necessary to understand many applied research results.

Again, judicious use of these types of graphical displays with clear labels and concrete explanations may be the only way to educate the public on substance and increase graphicacy levels simultaneously. I have placed occasional relational graphs in briefings or set them in prominent boxes with explanatory text in executive summaries of reports. In this way, sufficient explanations can be offered and individuals who wish to understand more about the substance of the presentation will be motivated to learn the relational form in the process. The unique strength of graphical displays can be called into play in this process—their atemporality. Viewers can return to the graph again and again when motivated to do so, to answer questions as they arise. When the graph—including labels and legends—is complete and can stand alone, the viewer can return and glean more information from it.

EXERCISES

1. Reexamine the graphs collected from popular publications for relational graphs. Are any relational? Do they use continuous variables?

2. Search through three or four volumes of a professional journal in your field that publishes quantitative research results. How many scatterplots are included? Do the plots complement the analysis presented in the text? Do the plots stand alone in terms of variable labels and a description of the units from which the batch of data were drawn? Critique the graph based on the suggestions in the last chapter.

3. Plot two variables from the data contained in Table 4.1 that you expect to be asymmetric. Now plot two variables that you expect may be "causally" related. Enter the data in a program that you will use for graphing and have the program plot the data. How do these plots differ from your original plots? Would the plots produced by the default options be easily understood by a nontechnical audience? Try to enhance their usefulness by customizing the program's output.

7

Graphical Complements for T-Tests and Analysis of Variance

Much of the work of applied research is directed at the study of variability of data. Program effects vary from one participant to the next; medical treatments vary in their effectiveness; public policies vary in their support and consequences; the implementation of interventions varies from one site to the next. Certainly, the prominent use of statistics in the fields of applied research has imprinted the idea of variability on the modern mind, but we have come to rely principally on summary statistics, which often fail to convey the variability in the data. As Cleveland (1985) has noted, "Numerical data analytical techniques—such as means, standard deviations, correlation coefficients, and *t*-tests—are essentially data reduction techniques" (p. 9). These commonly used summaries focus our attention on an estimate or on the results of a test of significance. They provide only a small amount of information from the batch of data being analyzed. Although we certainly grasp the idea of variability, our statistical techniques may obscure it, and as a result our audiences are often less well informed.

Perhaps nowhere has the tendency to reduce research findings to technical artifacts of statistical procedures and the criticism of this tendency been more prominent than in the analysis of experimental data. The field of evaluation, traditionally relying on the logic of experimental and quasi-experimental designs, is replete with criticisms of the dogmatic reliance on tests of significance. These criticisms have sometimes been misinterpreted (or misdirected) as critiques of the experimental tradition. I believe they are correctly focused at the reliance on statistical artifacts, which is an exceedingly reductionist tactic. The two applied social scientists who have codified evaluation as a form of inquiry, Cook and Campbell (1979), prefer to use "bounded magnitude estimates" to present the range of effects that could be expected for a fixed proportion of the cases (p. 41). Essentially, bounded magnitude estimates provide a concrete sense of the expected effect size, much like a confidence interval drawn around a point estimate. Bounded magnitude estimates provide much more information than point estimates or tests of significance.

The central problem is that the test of significance or the reporting of the effect size alone can be misleading (Reichardt & Gollob, 1989). In some cases, an effect can be statistically significant but meaningless from a practical standpoint; in other cases, the effect size estimate could appear to be very important but still be judged statistically insignificant. One solution to this problem is to use confidence intervals or plausibility brackets, which include the impact of identifiable threats to validity (Reichardt & Gollob, 1989). These techniques provide more information to the audience for the results, but graphical displays can go further, revealing the data and imprinting an image of the location and spread of the data. Viewers can see the discrepancies and the variations for themselves and make their own judgments. Of course, the graphs should be considered a complement to other statistical information and used to aid the decision-making process. Graphs can aid the analyst in better understanding statistical summaries and provide the audience with a visual depiction of the results free of the technical jargon of narrative explanations.

The principal analytical techniques used in experimental design and designs that approximate experiments are *t*-tests and analysis of variance (ANOVA). Certainly, a variety of more sophisticated techniques is being utilized to structure the data and allow for its complexities and deviations from the experimental ideal. Some of these designs involve longitudinal data, which can be graphed using the time series techniques in Chapter 5. In the remainder of this chapter, I concentrate on graphical methods that allow applied researchers to offer their audiences more access to data where the primary objective is to show comparisons between two or more groups. These types of graphs are often referred to as categorical plots, or graphs that plot a continuous variable by a grouping variable.

Both *t*-tests and ANOVA are complemented by the use of categorical plots, with the categorical plot used for ANOVA expanding the number of categories, as one of the advantages of ANOVA is the use of multiple groups. A graphical complement to ANOVA allows us to see the differences between means and distributions of responses for each group. This gives more information than the commonly cited statistical significance of the group differences and therefore overcomes some of the criticism directed at the use of statistical significance rather than the magnitude of difference. Graphs show both. Graphs also give meaning to the notion that the amount of variation explained in the dependent variable is due to the treatment, or independent variable. I begin with some graphical complements to *t*-tests.

T-TEST GRAPHICS

The simplest graphical form for presenting data appropriate for *t*-tests is a two-group version of the dot chart or another form of the histogram, such as the one shown in Figure 7.1. On this figure, dot histograms for both the treatment and control groups have been plotted vertically. It appears from the comparison of the distribution of the effects for both groups that the treatment registered a slight, positive effect. Although this graph shows the two distributions, it does not indicate the location

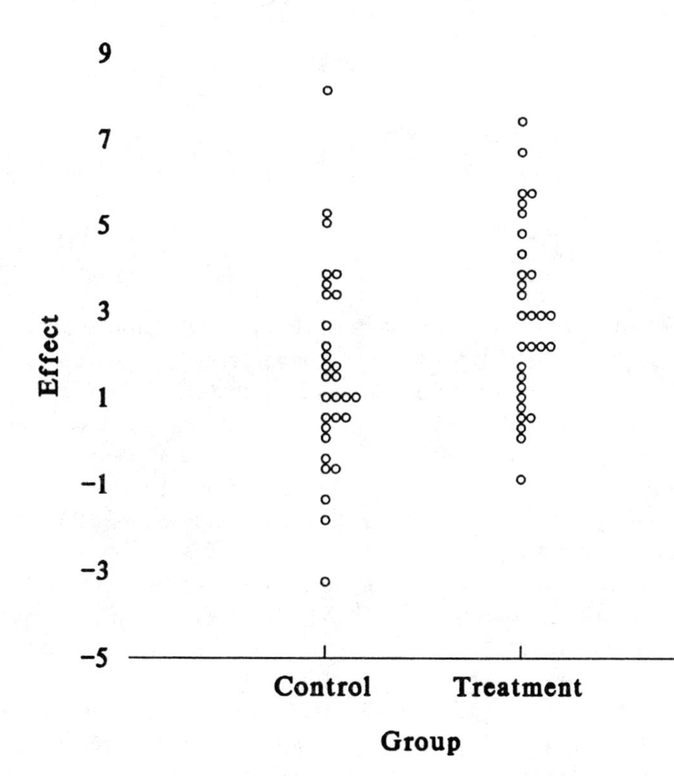

Figure 7.1. Comparison of Outcomes

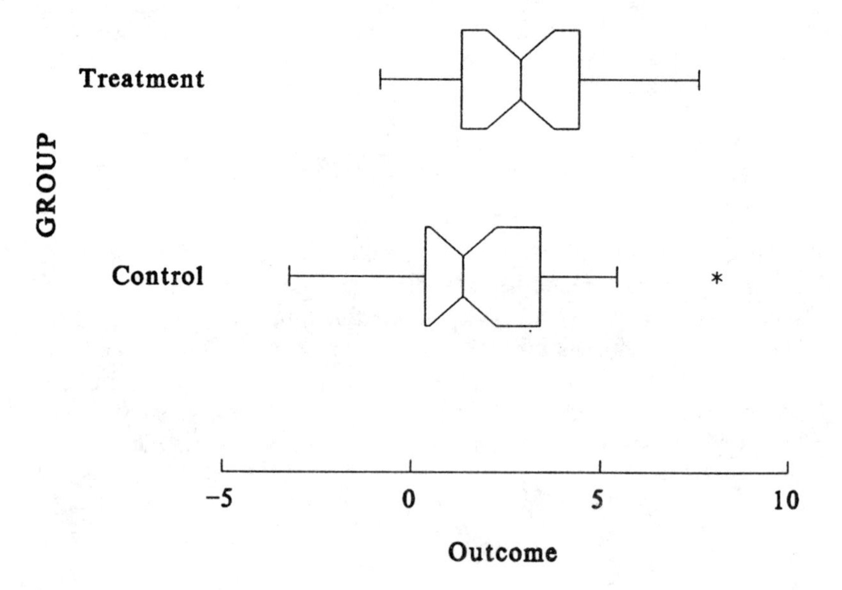

Figure 7.2. Comparison of Outcomes (Box Plot)

or center of the distribution, nor the *t*-test results. We observe the variability, but have no additional cues to determining whether our impression is accurate.

Box plots can be used for *t*-test data to show salient features of the distributions of the two groups being tested for significance. The notched box plot (see Figure 7.2) provides a visual test of significance (Benjamini, 1988; McGill, Tukey, & Larsen, 1978). Although not the *t*-test of means that we are most used to seeing, the graph does show a test of the difference between the median of the two groups. If the notches of two groups do not overlap, about 95 times out of 100, the population *medians* are different. These figures require specialized programming available in several statistical software programs with graphing capabilities, such as SAS, SPSS, and SYSTAT. To plot box plots for two or more groups, the categorical variable must be used to specify which groups are to be plotted. In Figure 7.2, the median effects do not appear to be statistically significant.

Often applied researchers wish to explore the difference between two groups on ordinal data or truncated continuous variables. In some cases, graphical displays such as those in the last two figures reveal little about

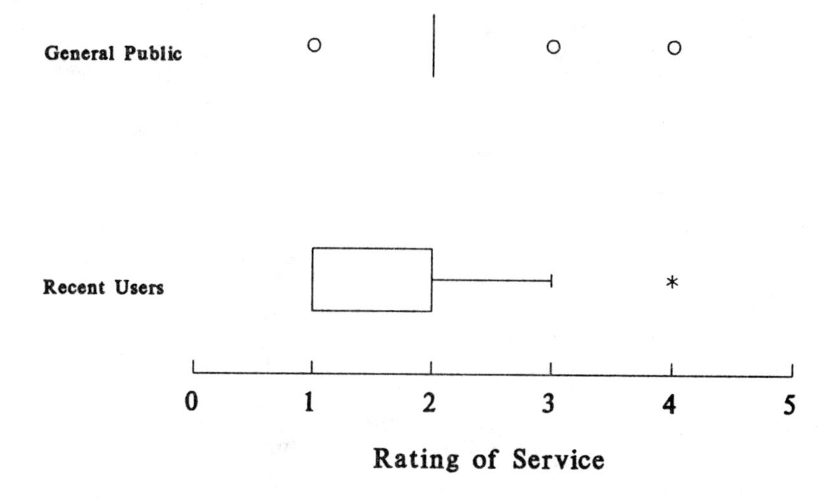

Figure 7.3. Rating Postal Service (Box Plot)

the data, as we see in Figure 7.3. In Figure 7.3, we see the responses of two groups to a question concerning the quality of postal services. One group is the general public and the other is recent users of postal services. The response categories consist of a 4-point continuum of nonspecific quantifiers, very good, good, fair, and poor, that are numerically identified as 1 to 4, respectively. The first plot is a box plot. Because some of the markers of the distribution fall on the same value, which obscures them from view, the plot is more difficult to read. Because the entire box for the general public appears on the good rating, it is evident that most of the ratings were good. The circles indicate that at least one individual rated the service in each of the other categories. Ratings for recent users are higher, as the box stretching between very good and good indicates; in this case the poor rating is an outlier, which is indicated in the plot by the * symbol.

In Figure 7.4, the means of the two groups are graphed with error bands around them to provide an indicator of location and a visual representation of statistical significance. The metric of the ratings has been changed to a 0 to 100 scale and the means calculated for both groups. Means are plotted by circles. The error bands are .5 of the denominator of the appropriate *t*-test for a specific probability of accepting a difference as significant when it is not, in fact, significant.

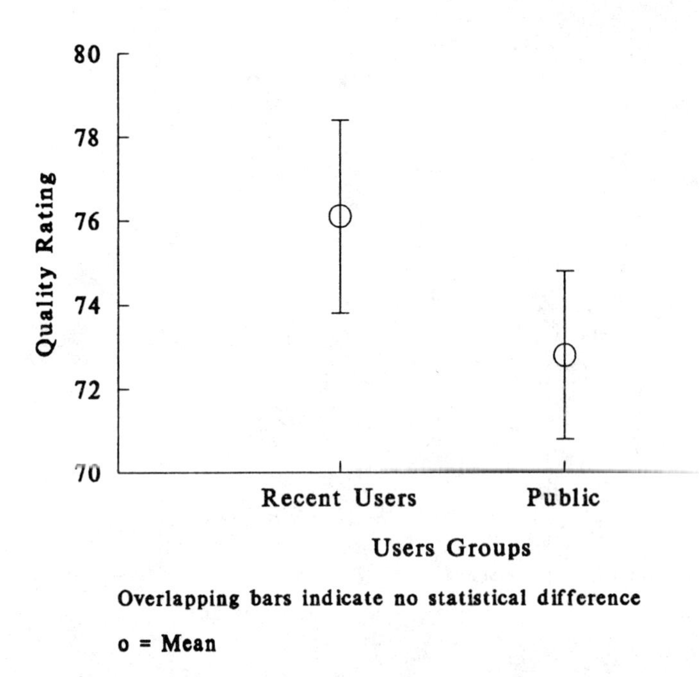

Figure 7.4. Rating Postal Service (Error Bar)

The overlapping bars indicate that the difference is not significant. This graphing technique can be used for paired *t*-tests as well as tests for independence, as is the case here. Obviously, the amount of data being graphed is more limited, but is especially useful for comparing the results of multiple *t*-tests. Although this graph does not provide some of the information contained in box plots, it is simple and provides both the magnitude of difference between the two groups and the statistical significance of the difference at a glance. This type of graph is also useful as a teaching tool about statistical significance.

A final *t*-test graph presents the raw data, with only the two groups used to structure the presentation of the values of the dependent variable (see Figure 7.5). The plot is a jittered categorical plot. Jittering assigns a small random number to each plot point that causes cases with similar values to be plotted in a clump rather than on top of one another. A line representing the mean has been added to show the location of both distributions. The difference between the two means is the numerator of the *t*-test statistic. This graph gives a visual impression of the

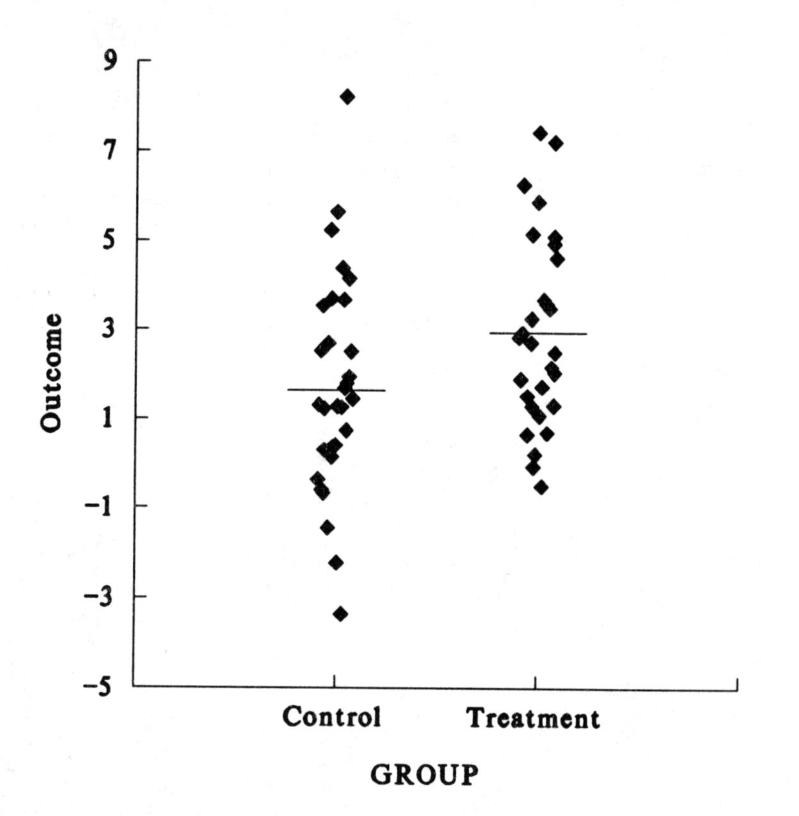

Figure 7.5. Experimental Outcomes

differences between two means in the context of the overall variability of the data. Also apparent in this plot is the number of data points in both groups. The number of cases is important in deciding whether the difference is likely due to chance.

The highlighted mean shows the audience the amount of difference in the means placed on the backdrop of the total variability of the two groups. It is hoped that this would stimulate viewers' thoughts about the impact of the treatment and the fact that there is substantial overlap between the two groups. A nontechnical audience might be assisted in grasping the idea that the impact of the treatment is not uniform, and although the impact is substantial for the overall group, improvement cannot be expected for each individual. In other words, all students will not benefit by an increase of 2.17 points on a standardized test score as the result of a new teaching method. The expected benefits lie in a range around the mean difference of 2.17. We may become attuned to picking up on whether the distribution appears to simply shift or whether the lower tail moves more or the upper end simply stretches out further. If an objective of the program is to raise the scores of lower performing individuals, then the way in which the distribution changes has implications for the desirability of the program.

ANALYSIS OF VARIANCE GRAPHICS

Analysis of variance (ANOVA) is used to test whether the differences that appear between different groups in terms of a response or dependent variable can reasonably be attributed to chance. A categorical variable distinguishes the groups. Therefore, most of the plots are categorical plots rather than Cartesian plots. Figure 7.6 shows the responses for all observations grouped by the categorical variable. The viewer can quickly find the mean response for each group and the distribution of responses. The mean for each group is indicated by the line in the column corresponding to that group. In this plot we have used a jitter technique to show the observations at each level of the response variable. Jittering simply locates an observation at a proximate point to its precise location based on a random process. Without jittering, the observations would have been plotted on top of each other at the precise intersection of the group and level variables. The jitter technique is especially useful with the variables that are being treated as continuous, but are measured by nonspecific quantifiers such as rating scales or other ordinal or truncated scales.

Statistical significance is not indicated in this graph, but viewers can examine the data for patterns and later test their own speculations. This plot was particularly interesting for policymakers presented with this data. The response variable was the level of commitment to the approach used in an educational performance monitoring system. The policymakers were being warned that teachers would not accept the new state system. But the data showed that the superintendents as a group were more concerned with the system than teachers—teachers, contrary to many opinions, were the most positive about the new system. The two other groups, school board members and representatives of educational interest groups, registered opinions between superintendents and teachers.

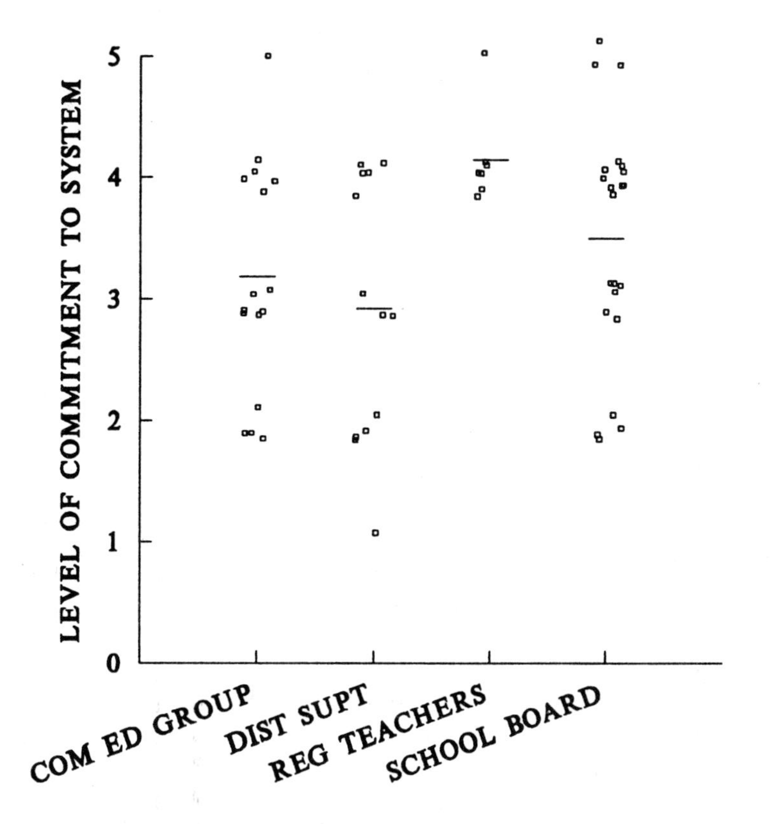

Figure 7.6. Rating Commitment to Indicator System

The jittered categorical plot shown in Figure 7.6 is actually two plots superimposed on each other. One is a plot of the group means; the other is a jittered categorical plot. To plot one on top of the other, a command sequence that includes the origin and the size of each axis is required. All labels, axes, and tick marks must be suppressed from one of the plots. Jittering can be accomplished by using a random number generator for values between .00 and .20, assigning a number for each variable of each unit to be plotted, and adding the variable and its random number to form a new variable. One statistical software program, SYSTAT, includes a command for performing this procedure without specialized programming. A grouped box plot, similar to those shown for *t*-tests, could also be used to complement an analysis of

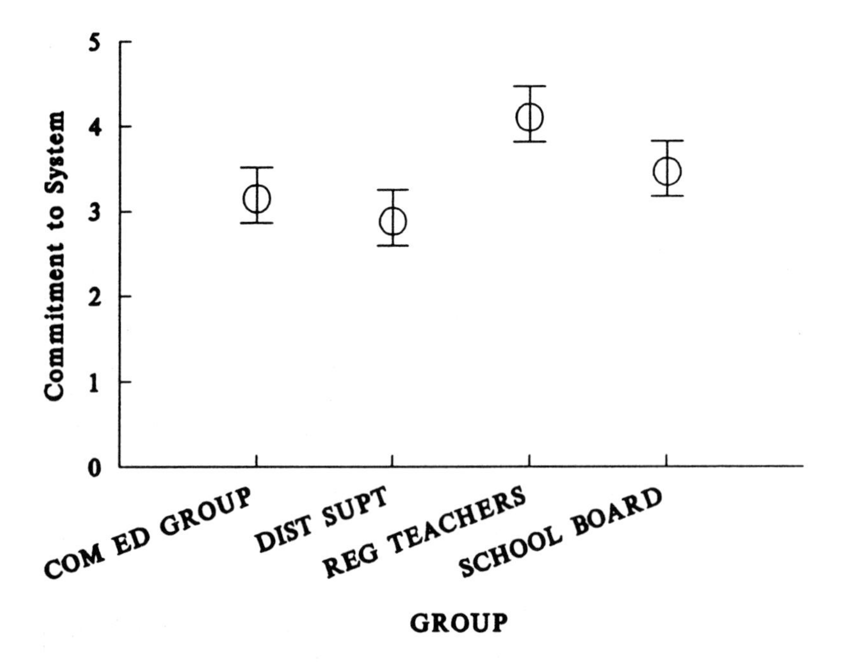

Figure 7.7. Rating Commitment To Indicator System

variance. The grouped box plot is more appropriate when the dependent variable is continuous.

The categorical plot of group means and error bars (see Figure 7.7) is another useful option. Means are once again represented by a straight line. The error bars are drawn at .5 of the critical distance computed by Scheffe's test. The attention of the audience can be focused on the group means and their differences in this graph, as the means are more readily visible than in the plot shown in Figure 7.6. Overlapping error bands indicate that two groups are too similar to be statistically different. Only teachers and district superintendents are statistically different.

For educators deciding whether the switch to a new teaching approach is justified by gains in student learning or for a physician deciding whether a different treatment modality should be used with patients, this graph could aid in decision making. The graph presents both the mean and the measure of variability most directly related to statistical significance, Scheffe's criterion. Quite possibly, this graph represents a reasonable compromise in the quest to give viewers enough information

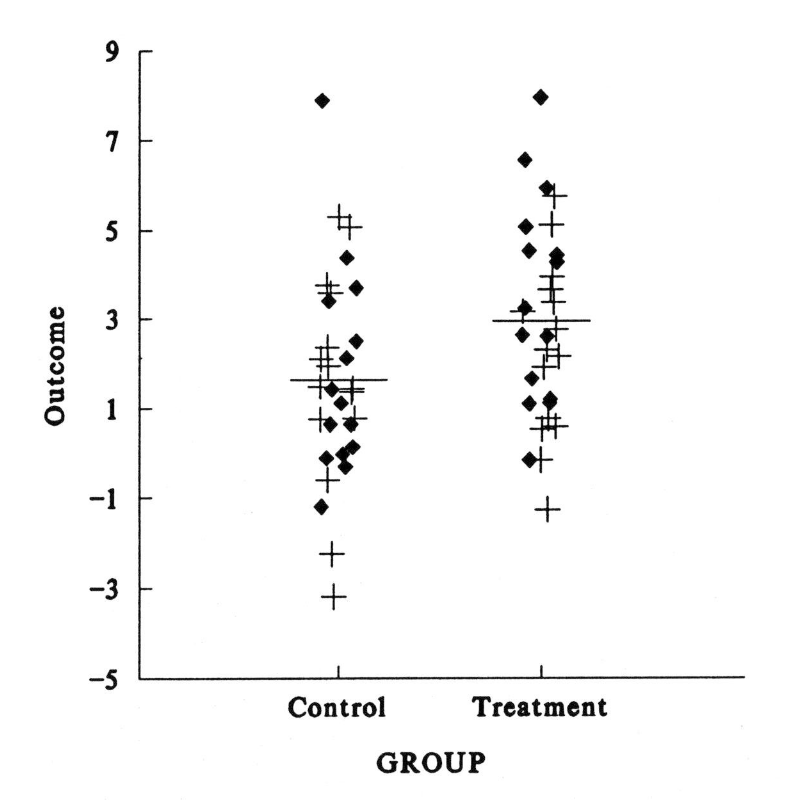

Figure 7.8. Experimental Outcomes

for coming to a conclusion without overwhelming them with data. Certainly, use of such graphs would reduce the focus on statistical significance as the *only* criterion for making a decision.

One additional graphic capacity is useful for the display of ANOVA data. Often ANOVA is used when two or more categorical variables are expected to have an impact on the dependent variable. The variables may be two aspects of the treatment regimen or another characteristic of the experiment, such as setting, or a characteristic of the participants, such as sex or race. Figure 7.8 illustrates this use of ANOVA with a graphical display that uses plus signs and diamonds as plot symbols to distinguish between two groups that are evenly distributed in the treatment and control group. In this case there is no difference other than chance between the treatment and control groups for the plus sign group, but the group represented by diamonds shows a large effect from the treatment. This graph aids in uncovering the underlying property of the experiment: The treatment has an effect on one group but not another. However, the effect of group membership is not statistically significant using ANOVA and therefore would have been missed without the graphical representation.

SUMMARY

T-tests and ANOVA are techniques that simplify the process of making judgments about data. They are both useful in determining whether differences exist between groups. However, both techniques boil down the amount of information to an extent that potential users of the data may not have sufficient information to act. They know that 95 out of 100 times differences as large as the one that occurs in the data would not be present if there were no true difference. Or alternatively, they know that the amount of the difference may not be large enough to rule out chance as the explanation. But because the size of the sample is one of a number of intervening variables, the use of statistical significance alone is not a sufficient guide to making a judgment.

Graphs can supplement tests of statistical significance and estimates of effect size. Graphical displays of data can show both the effect size differences and the variability of the dependent variable. They go beyond bounded magnitude estimates and plausibility brackets in the amount of information that can be presented to the audience. They cannot parcel out the impacts of the various threats to validity that are being adjusted through plausibility brackets (though the results of these adjustments could be graphically rather than narratively displayed), but they do provide a concrete look at the variation that underlies the dependent variable. Thus graphical displays create a more realistic view of the conditions that actually exist and another important point for making judgments.

Some of the graphical formats presented in this chapter are more useful for applied researchers to share with each other and for their own analysis than to present to lay audiences. The same is true for some of the graphs in the previous chapter. Other formats are shown that represent compromises between analytical graphics and descriptive graphics and may be appropriate for lay audiences. Too much information can be as stultifying as too little. The choice of a graphical display type and the modifications of some of the basic types shown in these two chapters often depends on the familiarity of viewers with graphical displays and their interest in analyzing the data for themselves. This, as we will discuss in the final chapter, involves tailoring displays to the audience in the process of generating competent graphical displays.

EXERCISES

1. Review the graphs that you have collected earlier. Are any of the graphs intended to show the differences in two or more groups? Do any of the graphs give an indication of the variability in the data? How could these graphs be improved?

2. Using the graphing software that you have available, design a box plot. Can you use a grouping variable to plot two or more box plots on the same graph? If not, select one group at a time and graph the box plot. NOTE: scales should be consistent and aligned.

3. Attempt a categorical plot, using the software or by hand. How would the plot be enhanced by jittering or error bars? Plot the error bars by hand after calculating their length.

8

Achieving Graphical Competence

A good graph looks stone simple. Simplicity permits viewers to go directly to the information, to view it, to see how it stacks up against their expectations, and if the graph is truly successful, to wonder. "All credibility, all good conscience, all evidence of truth come only from the senses," Nietzsche has told us. Graphical displays of data allow us to muster vision in pursuit of credibility and truth. In *A Natural History of the Senses,* Diane Ackerman (1991) reminds us of the dominance of vision:

> for us the world becomes most densely informative, most luscious, when we take it in through our eyes. It may even be that abstract thinking evolved from our eyes' elaborate struggle to make sense of what they saw. Seventy percent of the body's sense receptors cluster in the eyes, and it is mainly through seeing the world that we appraise it and understand it. (p. 230)

Seeing for ourselves rather than being told produces an understanding of the subject that approaches that of the researcher. Using graphical displays, we are no longer passive observers, we become active seekers of information. Depending upon time and interest, we understand some of the complexity of the subject under study, and perhaps, the limits of current knowledge.

Why haven't graphs produced this enlightenment and understanding?

For one thing, the practice of producing competent graphics is deceptively complex. A competent graphical display is transparent to the viewer: the display goes unnoticed and the data are attended to. As I have shown in the previous chapters, even an interesting batch of data run through a standard graphics package will not automatically result in a competent display. The outcome is no different from what one must expect from bludgeoning a batch of data with a statistical package. Training, patience, and a goal must come together to produce a useful, competent graph. The rarity of this confluence is indicated by the fact that a book showing *in*competent graphical displays would run several volumes.

Another factor is that graphics have been underutilized. I use the term underutilized in two senses. First, we have not produced enough graphical data displays. The number of displays has been hampered by the cost and difficulty of production. Even today with graphics software packages that access the data bases used for statistical analysis and that offer menu-driven selections of options, producing a competent graph is time consuming. Guidance in making choices concerning the options has been so limited that the trial and error involved in graphical design can gobble time and wear down patience. A well-meaning attempt to use graphs often ends in the publication of a narrative with few graphs. The second sense in which graphs are underutilized is the amount of information that they contain. A bar chart with three or four bars showing proportions of the whole or a time series with so much data that the viewer is overwhelmed provides a weak view of the batch of data. In both these cases, the graphs are underutilizing the data.

Underutilizing data graphics has a deleterious effect on the prospect of using graphs to convey complexity and understanding through data that goes beyond a particular instance of underutilization. Graphicacy, just like any other skill, benefits from practice. Graphical formats are stored in long-term memory. The more frequently they are recalled to decode graphical information and the more sophisticated the formats that are retained in long-term memory, the more graphicacy skills will increase. Certainly, popular publications are replete with graphical displays. But they aim low in terms of the sophistication of formats and amount of information. They substitute color and cleverness for substance. Interest quickly wears thin when little information is conveyed. More important, viewers have not garnered something from the graph that would teach them to use a more sophisticated graph or provide them with the reward that would motivate them to expend the effort to approach a more abstract or difficult graph. Graphical comprehension skills remain stunted. Wainer (1992) suggests that our ability to read graphs is "hard-wired" and diminishes by viewing awkwardly designed graphs.

How can we use graphics to inform and increase the understanding of a subject?

This question leads back to the discussion of graphical competence. First, we must recognize the interactive nature of graphical design. The producer of the graph and the consumer of the graph have a relationship that relies exclusively on the graphical display. The purpose of the producer is to convey information. The objective of the consumer is to know more about the subject. The graph is the medium. Based on this perspective, we can offer a number of principles that can achieve these dual objectives.

PRINCIPLES FOR ACHIEVING GRAPHICAL COMPETENCE

1. Give Primacy to the Data

Show the data. The purpose of graphical data displays is to inform viewers. To inform, the displays must contain meaningful data and the display format must provide access to the data. Applied researchers, based upon their knowledge of the subject, have chosen data that they consider meaningful. This decision is often made in consultation with a client or research sponsor. Occasionally, the data collection effort will not yield the expected payoffs: Identified participants refuse to respond, the program to which the data collection is tied undergoes significant change, or the technical aspects of the measurement process fail to meet accepted criteria. For all of these reasons, the data collected may turn out to be less than meaningful, but for well-planned research conducted by experienced researchers, the presumption favors the data being considered meaningful. The data that applied researchers collect has been acquired at a cost of time and human effort. If it was worthy of collection, it is worthy of being displayed and analyzed.

In presenting meaningful data, one central idea should be held in mind: Comparisons are the means by which graphical information is digested. Graphicacy can be summed up in terms of the following comparisons:

> Level 1 graphicacy requires the ability to make a comparison between the position of the graphical symbol used to represent the data and the position or other graphical device that encodes the value of the data. For example, to read the values from a scatterplot, the viewer must be able to compare the point to its position along both axes. Level 1 graphicacy is the comparison between a specific piece of graphical information and the referent that is unique to the type of graphical display. An accurate description of the observation, developed through the comparison, is the objective of Level 1 graphicacy.
>
> Level 2 graphicacy is the ability to compare the position of multiple graphical symbols and discern a norm. The norm may be an average, a trend, or a slope. For a time series graph, the norm is the overall trend; for a bar chart, it is the average or point estimate. Level 2 graphicacy

assumes Level 1 skills. To construct a norm, the viewer must be able to interpret the position of all the graphical symbols that encode the data and form an image or overview of the norm. The comparison is across each of the graphical symbols in Level 2 graphicacy.

Level 3 graphicacy involves the ability to make comparisons between trends or between patterns and specific cases. In carrying out Level 3 graphicacy, the viewer goes from describing the subject to analyzing the data. The pattern of data in the 10 most recent time periods might be compared with the earlier trends using a time series graph. The comparison forces the question: Why?

Showing the data amounts to facilitating comparisons. The viewers should be able to retrieve specific data from the graph; they should be able to construct a norm; they should be able to analyze the data from the graph by comparing trends, patterns, and observations. Of course, not all graphs will place equal emphasis on each of the three types of exercises. The researcher must ask: What is the most meaningful, most important exercise for the audience to carry out with this data? One graph may not facilitate all the meaningful comparisons. As I noted in Chapter 5 in the context of multivariate time series, more than one graph may be needed to facilitate the multiple purposes that the applied researcher has as an objective. Attempting to do too much with a single graph may hide the data.

2. Design Information-Rich Graphics

Applied researchers should strive to produce graphics that reveal the underlying complexity of the subjects with which they deal. Graphs can show relationships that exist and dispel myths about relationships that do not. Graphical displays show patterns and exceptions to those patterns. Graphs give meaning to statistical summaries. They enhance the credibility of findings: seeing is believing, after all. But they only perform these functions when they contain the data necessary to carry them out.

In practice, carrying out these functions will require abstract graphical forms and high data densities. Relational graphics, the now-familiar Cartesian plot, are needed to show the relationship between two variables. Both variables can be individually described with simpler graphical forms, such as bar charts, but their covariance is most directly presented through the use of the two-variable plot. Graphs will often display large, multivariate batches of data. All other things being equal, greater data densities are preferred. But as we learn from economics, all things are never equal. Audiences less familiar with decoding graphical data may find the cost too dear to retrieve information from data dense graphical displays. A novel or unusual graphical display will also limit the optimal data densities. These conditions should not deter the use of data dense graphs. Rather they should be used by the researchers to assess what they are attempting to convey and tailor a design to the data *and* the audience. As the audience becomes more adroit with graphical displays, that is, as their skills with graphicacy increase, the data density can be increased.

Information-rich graphs necessitate the concentration of data ink and the removal of nondata ink. Applied researchers should pay special attention to the means used to encode data and attempt to encode as much as possible without creating an overload of symbols and nuance. Stripping away the conventions and superfluous items that do not display data is another important practice. Removing grid lines is often useful, especially when the primary purpose of the graph is to encourage higher levels of graphicacy. Eliminating the top and right axis lines in scatterplots removes ink that does not convey information. Another reduction of data ink that I encourage is removing the line that connects the points on time series graphs. However, many of my students object to this. They object principally on the grounds that the convention is so well established that they can't immediately interpret the data points as a time series. Once they get past the initial shock, however, they use their eyes more actively and perform higher level tasks at finer levels of distinction with the data. It is important to take the perspective of providing information-rich graphs with high concentrations of data ink into the design process.

3. Tailor Designs to Reveal Data

Taking the primary purpose of graphs as informing their audience, applied researchers should tailor designs to the data and the audience. Often this will involve many iterations of the display, with the researcher testing one improvement after another. And just as often, the improvement does not work out. Fortunately, three criteria can be used in the iterative process of graphical production to judge the success of a new graphical idea:

Avoid distortion
Enhance clarity
Encourage important comparisons

Each of these criteria should be applied as successive versions of the graph are produced.

Avoid Distortion. The size of the effect, be it indicated by a difference or a slope, should be equivalent to the size displayed in the graph. This phenomena is measured by the "lie factor," or the amount of distortion in the display of the effect (Tufte, 1983). Distortion can occur by a willful, malicious act, or by the actions of a graphical designer who values the medium more than the message, or, I suspect most commonly, by unintentional actions. Although applied researchers are as subject to human frailties as others, deliberate distortion is not the primary source of distortion. Graphical designers or graph makers unfamiliar with data graphics or the data can distort the data by using graphical software defaults or options inappropriate for the data. Pseudo 3-D pie charts and time series scale dimensions that result in overly tall or overly wide displays cause distortions.

Using size or color saturation to convey quantitative information can also cause distortion. Perception of the differences in size is a function of the actual difference in size. Unfortunately, the relationship is not one to one. Differences in size are perceived as less than the actual difference recorded in the graph. Therefore, even if the graph accurately conveys the difference, it is likely to be misperceived. This functional relationship applies to many phenomena and has been summarized as Stevens's power law. Perceptions of differences in lengths come closer to a one-to-one relationship than area or color saturation differences, for example. Using lengths such as bars and the distances from the axes in two-variable plots to convey most information seems to be safe.

One additional comment is in order. Sometimes it is not clear what the information-carrying symbol is. In a pictogram—a graphical device that uses a picture of an object to convey information—it can be unclear whether the height of the picture, the size (area) of the picture, or the volume of the picture provides the visual cue. Such confusion should be avoided by avoiding pictograms entirely. But confusion also arises when three-dimensional or even two-dimensional bars are used to convey information. Are the data represented by the height of the bar or its area? By convention, we expect it to be height, and it generally is. But bars should be thin to avoid this confusion.

Enhance Clarity. In tailoring designs, the importance of clarity can not be overstated. We must continually ask: What is the purpose of this graph? In the iterative process of tailoring the displays, it is easy to lose sight of the original purpose and whether the graph is achieving that purpose. Clarity in graphical displays requires the avoidance of ambiguity. Some potential sources of ambiguity have been discussed above, such as ambiguity about which physical aspects of the display convey the information—length, size, or volume? Here it is important to stress the importance of labels, titles, and legends.

A graphical display should be able to stand on its own. In addition to the data represented symbolically, there should be enough information for the audience to construct meaning from the display. Not enough technical information about the measures and the data collected can be provided to answer all questions, but it should be sufficient to decode the graph. Bertin (1967/1983) recommends that the title contain the "invariant," the thing that is common to all data represented in the graph. For a graph displaying parts of the whole, the whole should be in the title. For example, the pie chart in Figure 3.4 could be titled: "Federal Budget Allocations for 1953." It is useful to include the metric, in this case percentages, in the title or in the labels or legends.

For a scatterplot, the nature of the observations should be included in the title. This often involves a definition of the study population (Henry, 1990). For the scatterplots in Chapter 6, the title should include "School Districts in Virginia." For the plot that displays the relationship between the percentage of students eligible for free or reduced-price lunch and the percentage of first-grade students failing one or more entry tests, I would use the title, "Relationship Between Poverty and Failing Entry Tests in Virginia School Districts." The axes should be labeled with a descriptive variable label and the metric, in this case percentages. Some suggest using the title to convey the author's message. I find this a bit too leading and would rather stick to the facts and let the viewers develop their own conclusions.

Ambiguity can also occur in the labeling of axes and in the use of legends. Labels should be descriptive rather than technical. In the scatterplots in Chapter 6, the percentage of students eligible for free or reduced-price lunch is a measure of the level of poverty in the school district. In the title, I would convey the construct or the concept of the relationship that was being displayed. I would use a similar descriptive label for the horizontal axis with a lay audience. This practice will undoubtedly cause consternation among technical and subject matter specialists. For these audiences, the technical detail is required, but for a more general audience, including education policymakers, the descriptive label communicates more than the technical label. This choice is ultimately one of construct validity.

Legends are generally difficult to use. Legends needed to identify graphical symbols require viewers to shift their eyes back and forth as if they were watching a Ping-Pong match. The constant effort of using the legend can be a distraction, an inconvenience, and a source of error in retrieving the data. For two or three symbols, the use of legends is tolerable, but any more can cause a processing overload. Many graphic packages allow the labeling of the symbol within the display space, such as in the bar in Figure 1.2, or extended directly from the line as in Figure 5.12. When proximate labels are not possible, care should be taken to make the symbols as distinct as possible. Bertin (1967/1983) provides some advice on this, such as the use of symbols for dot (·), dash (–), and plus (+) as the most easily distinguished point symbols. Lines are more difficult to make distinct and areas may produce moire effects when hatching or cross-hatching is used. Figure 1.5 presents a desirable alternative for a cumulative time series graph. Important in shadings in this graph and the shadings used in the seventh panel of Figure 3.4 is to maximize the contrasts between adjacent areas. In any case, legends for graphical symbols require memorization, allow ambiguity to creep in, and slow down the decoding of information.

Encourage Important Comparisons. Important comparisons are encouraged by placing the items to be compared in close proximity. This is the concept behind grouping multiple bar charts. If the bars are grouped, we compare the responses within the groups most easily. If the bars are arranged by response categories, we make between-group comparisons most readily. An example of this is provided in Text Graph 3.4. Proximity is one key to encouraging comparisons, but other considerations enter into the process.

Displaying data on identical scales is essential if comparisons are to be accurately made. The scaling question enters in several ways. For time series data, the intervals in the display must be equivalent. As a counterexample, Figure 5.5 uses unequal time intervals that invalidate comparisons: Specifically, the first four intervals are for 10 years and the final one for 8. Equally important is the use of identical scales on every graph that is to be compared. This sometimes causes less than optimal scaling on individual graphs, but the data are directly comparable. The plot matrix in Figure 6.3 is an example of good practice in this regard. Occasionally, the comparability of displays can be questioned in a single graph when a transformation is used that affects the original metric, such as log scale graphs. Truth in labeling, as in Figure 5.13c, is the only antidote for this problem.

4. Remember the Audience

Early in this chapter I restated the presumption that permeates the entire book: achieving competent graphical displays is an interactive process. Viewers are active participants in the retrieval of information and it is the unique characteristic of graphical data displays, that is, atemporality, that provides them access to the information on their own terms. Graphical data cannot be force fed. Graphs are the medium of a free exchange of information. Data graphers are seeking to stimulate the demand for their product by the audience. This must be done carefully.

Care must be taken to respect the current level of graphicacy of the audience and constantly seek to foster greater levels of graphicacy and greater comfort with data. Respect for the audience is paramount in graphical display. As Tufte (1983, pp. 80-81) has noted, we must consider the viewer of our graphs to be as intelligent as ourselves. A cynical or condescending view of the audience taints the graphical design process and leads to failed graphs. Graphs developed under these circumstances are weightless, without heft or substance. We have a fact, but just a fact.

It is not necessary to present graphical iterations to the audience, but it may be useful to begin with a familiar format and then move to a more challenging one to develop graphicacy with increasingly sophisticated formats. It may be useful to label a point or provide a text box below the graph to acquaint the audience with the way in which the data are coded. The legend for Figure 4.9 was quite specific and in the original report it was accompanied by a concrete example that showed how the information could be decoded. When the format for a graphic is unique and different, it requires some explanation. Abstract displays such as relational graphs generally require more explanation until the formats are comfortably recorded in long-term memory.

AESTHETICS AND DEMOCRACY

Graphical displays are in one sense aesthetic expressions of a process that places little intrinsic value on aesthetics. Truth, accuracy, and efficiency are the criteria of science that we impose on graphical design. And well we should. But the opportunities for elegance and visual appeal from graceful proportions and appealing type fonts should not be lost. Color, properly added, lends an appeal to graphical display. For an exploration of graphics with exceptional visual appeal, leaf through Tufte's (1990) *Envisioning Information.* For me, more than anything else, that book gives a heightened awareness of graphical aesthetics. It lays open the art of data graphing. It is a clear departure from the craft orientation of his (1983) *The Visual Display of Quantitative Information,* which has been so often referenced in this text. Both press home the notion of the importance of aesthetics. These books are art histories for quantitative data display.

Awareness and interest in graphical aesthetics is its own virtue and some of us will take pleasure in improving the aesthetics of our graphs and the appreciation of the work of others. But aesthetics is not without its practical value. The production quality of graphics is increasing rapidly with the use of color printers and professional firms that concentrate on graphic production for well-funded private sector projects. The audience for whose attention applied research data must compete is becoming accustomed to these types of graphics. Professional graphics in popular publications and public relations reports set the standards. The slick graphs in annual reports to stockholders—soon to be required in mutual fund reports—exhibit a production quality that social science seldom measures up to. Of course, they often don't meet the standards of accuracy and efficiency set forth in this text, but that's not the point. A public accustomed to bold graphs may overlook important data conveyed in shades of gray. Improving the aesthetics of data graphs will become an issue for applied researchers who want lay audiences to understand their work.

This point raises the other value lurking behind the use of competent data graphics. Graphical displays of data have implications for empowerment and democracy. Graphical displays that reveal data, that allow citizens to understand the complex relationships that undergird the operation of public programs and a wide range of interventions, can encourage questions and actions to address problems. With information goes the potential to use the information. School performance data is one example. How well are schools, school districts, or state school systems performing? Should parents, as guardians of the interests of their children, and taxpayers have that information? Is it impossible to show performance comparisons of schools or school districts with students of similar socioeconomic levels and similar community characteristics? Would this information be useful to parents in making decisions about where they will live and the choice of schools? If parents and taxpayers have information, can they impact the delivery of educational services?

Whether the audience for educational performance data would follow the course sketched out in these questions is uncertain. Currently, few have access to the data and the data formats for the attempts at providing access (e.g., the U.S. Secretary of Education's Wall Chart) have obscured rather than encouraged comparisons. Clearly the potential for empowerment is there when access to data is opened up through competent graphical displays. MacDonald (1976) has stimulated the concern for the role of evaluator and applied researcher within a democracy as information broker. Competent graphical displays of data will aid in carrying out this role in the future.

References

Ackerman, D. (1991). *A natural history of the senses.* New York: Random House.

Anderson, E. (1957). Semigraphical method for the analysis of complex problems. *Proceedings of the National Academy of Science, 13,* 923-927.

Atlanta Journal-Constitution/Georgia State University Poll (1993, Spring). Applied Research Center, Georgia State University.

Baird, J. C. (1970). *Psychophysical Analysis of Visual Space.* New York: Pergamon.

Behrens, J. T., & Stock, W. A. (1989, January). *Perceptual illusions in box-plots: Judgments in length.* Paper presented at the meeting of the American Statistical Association, San Diego.

Behrens, J. T., Stock, W. A., & Sedgwick, C. A. (1990). Judgment errors in elementary box plot displays. *Communications in Statistics, 19,* 245-262.

Beniger, J. R., & Robyn, D. L. (1978). Quantitative graphics in statistics: A brief history. *American Statistician, 32*(1), 1-11.

Benjamini, Y. (1988). Opening the box of a boxplot. *American Statistician, 42*(4), 257-262.

Bennett, J. H. (Ed.). (1971-1974). *Collected papers of R. A. Fisher.* (Vols. 1-5). Adelaide: University of Adelaide.

Berne, R., & Stiefel, L. (1984). *The measurement of equity in school finance: Conceptual, methodological, and empirical dimensions.* Baltimore, MD: Johns Hopkins University Press.

Bertin, J. (1983). *Semiology of graphics: Diagrams networks maps* (W. J. Berg, Trans.). Madison, Wisconsin: University of Wisconsin Press. (Original work published 1967)

Britain this week. (1992, March). *The Economist,* pp. 67-70.

Chambers, J. M., Cleveland, W. S., Kleiner, B., & Tukey, P. A. (1983). *Graphical methods for data analysis.* Boston: Duxbury Press.

Chernoff, H. (1973). The use of faces to represent points in *k*-dimensional space graphically. *Journal of the American Statistical Association, 68,* 361-368.

Cleveland, W. S. (1979). Robust locally weighted regression and smoothing scatterplots. *Journal of the American Statistical Association, 74,* 829-836.

Cleveland, W. S. (1984). Graphical methods for data presentation: Full scale breaks, dot charts, and multibased logging. *American Statistician, 38*(4), 270-280.

Cleveland, W. S. (1985). *The elements of graphing data.* Monterey, CA: Wadsworth Advanced Books.

Cleveland, W. S. (1987). Research in statistical graphics. *Journal of the American Statistical Association, 82*(398), 419-421.

Cleveland, W. S., Diaconis, P., & McGill, R. (1982). Variables on scatterplots look more highly correlated when the scales are increased. *Science, 216,* 1138-1141.

Cleveland, W. S., Harris, C. S., & McGill, R. (1983). Experiments on quantitative judgements of graphs and maps. *Bell System Technical Journal,* 62(6), 1659.

Cleveland, W. S., & McGill, R. (1984). Graphical perception: Theory, experimentation, and application to the development of graphical methods. *Journal of the American Statistical Association, 79,* 531-554.

Cleveland, W. S., & McGill, R. (1985). The many faces of a scatterplot. *Journal of the American Statistical Association, 79,* 807-822.

Cook, T. D., & Campbell, D. T. (1979). *Quasi-experimentation: Design and analysis issues for field settings.* Boston: Houghton Mifflin.

Eells, W. C. (1926). The relative merits of circles and bars for representing component parts. *Journal of the American Statistical Association, 21,* 119-132.

Ericsson, K. A., Chase, W. G., & Faloon, S. (1980). Acquisition of a memory skill. *Science, 208,* 1181-1182.

FBI. (1991). *Uniform crime reports for the United States.* Washington, DC: Federal Bureau of Investigation.

Fienburg, S. E. (1979). Graphical methods in statistics. *American Statistician, 33*(4), 165-178.

Funkhouser, G. H. (1936, January). A note on a tenth century graph. *Osiris,* pp. 260-262.

Ginsburg, A. L., Noel, J., & Plisko, V. W. (1988). Lessons from the wall chart. *Education Evaluation and Policy Analysis, 10*(1), 1-12.

Hedrick, T. E., Bickman, L., & Rog, D. J. (1993). *Applied research design: A practical guide.* Newbury Park, CA: Sage.

Henry, G. T. (1990). *Practical sampling.* Newbury Park, CA: Sage.

Henry, G. T. (1993). Using graphical displays for evaluation data. *Evaluation Review, 17*(1), 60-78.

Henry, G. T., McTaggart, M. J., & McMillan, J. H. (1992). Establishing benchmarks for outcome indicators: A statistical approach to developing performance standards. *Evaluation Review, 16*(2), 131-150.

Jarvenpaa, S. L., & Dickson, G. W. (1988). Graphics and managerial decision making: Research based guidelines. *Communications of the ACM, 31*(6), 764-774.

Joint Legislative Audit and Review Commission. (1991). *Review of Virginia's Parole Process* (Senate Document No. 4). Richmond, VA: Virginia General Assembly Bill Room.

Joint Legislative Audit and Review Commission. (1992). *Review of the Virginia Medicaid Program* (Senate Document No. 27). Richmond, VA: Virginia General Assembly Bill Room.

Kosslyn, S. (1985). Graphics and human information processing: A review of five books. *Journal of the American Statistical Association, 80*(391), 499-512.

MacDonald, J. B. (1976). Evaluation and the control of education. In D. Tawney (Ed.), *Curriculum evaluation today: Trends and implications* (pp. 123-136). School Council Research Studies. London: Macmillan.

Marris, P. (1992, May). How social research could inform debate over urban problems. *Chronicle of Higher Education,* p. A40.

McGill, R., Tukey, J. W., & Larsen, W. A. (1978). Variations of box plots. *The American Statistician, 32,* 12-16.

Playfair, W. (1786). *Commercial and Political Atlas.* London.

Poister, T. H., & Henry, G. T. (1994). Citizen ratings of public and private service quality: A comparative perspective. *Public Administration Review, 54*(2), 155-160.

Powell, B., & Steelman, L. C. (1984) Variations in state SAT performance: Meaningful or misleading? *Harvard Educational Review, 54,* 389-412.

Powell, B., & Steelman, L. C. (1987). On state SAT research: A response to Wainer. *Journal of Educational Measurement, 24*(1), 84-89.

Reichardt, C. S., & Gollob, H. F. (1989). *How research practice can be improved by using confidence intervals instead of hypothesis tests.* Unpublished manuscript, Department of Psychology, University of Denver, Denver, CO.

Rossi, P. H., & Freeman, H. E. (1993). *Evaluation: A systematic approach* (5th ed.). Newbury Park, CA: Sage.

Schmid, C. F., & Schmid, S. E. (1979). *Handbook of graphic presentation* (2nd ed.). New York: John Wiley.

Simkin, D., & Hastie, R. (1987). An information-processing analysis of graph perception. *Journal of the American Statistical Association, 82* (398), 454-465.

Stock, W. A., & Behrens, J. T. (1991). Box, line, and midgap plots: Effects of display characteristics on the accuracy and bias of estimates of whisker lengths. *Journal of Educational Statistics, 16*(1), 1-20.

Tufte, E. R. (1983). *The visual display of quantitative information.* Cheshire, CT: Graphics Press.

Tufte, E. R. (1990). *Envisioning information.* Cheshire, CT: Graphic Press.

Tukey, J. W. (1977). *Exploratory data analysis.* Reading, MA: Addison-Wesley.

Tukey, J. W. (1988). Some graphic and semigraphic displays. In W. S. Cleveland (Ed.), *The collected works of John W. Tukey* (pp. 37-62). Pacific Grove, CA: Wadsworth & Brooks.

U.S. Bureau of the Census. (1993). *Statistical Abstracts of the United States.* Washington, DC: U.S. Government Printing Office.

U.S. General Accounting Office. (1992). *Budget policy: Prompt action necessary to avert long-term damage to the economy* (GAO/OCG-92-2). Washington, DC: U.S. Government Printing Office.

U.S. General Accounting Office. (1993). *Budget issues: A comparison of fiscal year 1992 budget estimates and actual results* (GAO/AFMD-93-51). Washington, DC: U.S. Government Printing Office.

Virginia Department of Education. (1988). *Teacher salary survey.* Richmond, VA: Author.

Virginia Department of Education. (1993). *Outcome accountability project: 1993 Virginia summary report.* Richmond, VA: Author.

Wainer, H. (1986). Five pitfalls encountered while trying to compare states on their SAT scores. *Journal of Educational Measurement, 23*(1), 69-81.

Wainer, H. (1991). The isthmus of acceptance: A graphical tool for function-based item analysis and test construction. *Journal of Educational Statistics, 16*(2), 109-124.

Wainer, H. (1992). Understanding graphs and tables. *Educational Researcher, 21*(1), 14-23.

Wheelwright, S. C., & Makridakis, S. (1980). *Forecasting methods for management* (3rd ed.). New York: John Wiley.

Wilkinson, L. (1990). *SYGRAPH* [computer program]. Evanston, IL: SYSTAT.

Index

About the Author

Gary T. Henry directs the Applied Research Center at Georgia State University. He serves on the faculty of the School of Public Administration and Urban Studies and the Department of Political Science. Currently, he is the principal investigator for a research and development project that will produce indicators of student performance for schools and school systems in Georgia. He serves as Associate Editor of *New Directions in Program Evaluation* and is active in the American Evaluation Association's topical interest group in Theory and Quantitative Methods. He is a frequent contributor to *Evaluation Review* and *Public Administration Review.* He is currently writing about policy knowledge and the relationship among applied research, policy knowledge, and democratic political processes.